EDINBURGH & GLASGOW RAILWAY GUIDEBOOK

DON MARTIN &
A. A. MACLEAN

STRATHKELVIN
DISTRICT
for Leisure

AULD KIRK MUSEUM PUBLICATIONS No. 20

ISBN 0 904966 34 8

Published by
Strathkelvin District Libraries & Museums
170 Kirkintilloch Road, Bishopbriggs, Glasgow
G64 2LX

Printed by
Cordfall Ltd
041 332 4640

Contents

Foreword 4

The Railway Today 7

Down the Years 32

History from the Train 62

Stepping out from the Stations 71

Backward Glances 91

Recent Events 111

Today's Trains 116

Mile by Mile 118

Bibliography 126

Acknowledgements 128

Foreword

When the Centenary of the opening of a railway line between Scotland's two principal cities by the Edinburgh & Glasgow Railway Company came around in 1942, Britain was covered by the dark clouds of war. Needless to say the weighty matters of wartime survival far outweighed any ideas of celebration. The line itself – referred to in the Scotsman newspaper at its opening as "This magnificent line of railway" – was being busily worked not just carrying passengers, but also freight and military equipment essential in maintaining the war effort. It is therefore doubly important that the 150th anniversary should be marked and it is equally appropriate that a book celebrating the line and its surrounding environment should be launched to coincide with the Sesquicentenary. I am very grateful to Strathkelvin District Council for giving me this chance to say a few words regarding this line which I and my railway colleagues take a great pride in running today.

By the time this guidebook is being read by you – and hopefully many, many others – passengers will have been able to enjoy travelling between Scotland's two major cities by train for one and a half centuries. The building and opening of this railway on 18th February 1842 was a significant achievement as it was the first line in Scotland to connect two major cities, and the importance of the route has not declined with the passage of time and can today boast to being ScotRail's flagship route.

Today there is the strange daily mass movement of Glaswegians to Edinburgh and vice versa as workers who live at one end of the line travel to work at the other. Then in the evening they all return from whence they came. It is all the more strange when you consider it ties together two cities which have enjoyed a healthy rivalry for generations. The many important places the line serves in between Glasgow and Edinburgh – Bishopbriggs, Lenzie, Croy, Falkirk High, Polmont, Linlithgow and Haymarket – and their role in deciding the route the line follows, should not be ignored. In no way should the line be seen merely connecting the two larger cities. As much today as in the early days, these smaller towns and villages are vital to the existence of the line both in commercial terms and passenger needs. Today ScotRail are proud to carry on the traditions of the last 150 years in providing a fast, efficient, modern railway for both commuters and leisure travellers on the line between Edinburgh and Glasgow.

At present we operate a dedicated fleet of ScotRail's latest Class 158 Express trains on a half-hourly departure time on the hour and half-hour from both cities. These trains combine speed with comfort and strength with light weight, giving passengers an enjoyable 50-minute journey time between the two conurbations. The 158s boast a high standard of passenger accommodation with air-conditioning, full carpeting, double-glazing and comfortable seats with the most generous leg-room on any BR standard class vehicle. A far cry from the steam-hauled wooden carriages which took two hours to make the same jouney, including having to be winch hauled up the long incline from Queen Street to Cowlairs. We at ScotRail are very proud to continue running services over this prestigious route. Our aim is simple – to provide a railway of excellence of which all Scotland can be proud to have and to use.

With this publication people can easily see the range of leisure pursuits and places of interest which can be reached from the "E & G" route. E & G is the term still affectionately applied to the line by railwaymen despite the demise of the Edinburgh & Glasgow Railway Company – from where the initials come – around 125 years ago. I wish the publication every success and hope that in another 150 years the head of the railways in Scotland is able to write in a similar vein of how proud people are of 300 years successful running of the E & G.

We recall railways were built as a commercial profitable venture. This book recalls those who pioneered the route with their labour. Today its continued success will depend on those dedicated railway men and women serving the passengers who travel on Scotland's railways.

Cyril Bleasdale

Cyril Bleasdale,
Director ScotRail.
18th February 1992.

edinburgh
Shuttle
6
EXPRESS
38 times a day
every half hour.
6

The Railway Today

This chapter describes the railway infrastructure only. Historical and topographical features are described in the 'History from the Train' and 'Stepping out from the Stations' sections. Despite the preference of the original Edinburgh & Glasgow Railway Company to the contrary, and that of the first guidebook to the line, written by John Willox in 1842, the various features are described in a West-East direction. It is recommended that at Queen Street Station, Glasgow, travellers find for themselves a comfortable window seat on a Class 158 Glasgow-Edinburgh diesel train, facing the direction of travel. A seat on the left-hand side of the train is much to be preferred, since there are many more features worth noting, both of railway and general interest, on the 'up' or north side of the line. The route described is the direct one from Glasgow Queen Street to Edinburgh Waverley via Falkirk High. The loop line from Greenhill Junction to Polmont, via Falkirk Grahamston, is described in a separate section at the end of the chapter.

Glasgow to Greenhill

GLASGOW (QUEEN STREET)

Our journey begins in the loftily arched train shed of Glasgow (Queen Street) High Level Station. Although this terminus was built for the opening of the railway in 1842, the magnificent tiered glass roof dates from the period 1877-82, when the station was extensively reconstructed. It was designed by James Carswell, Civil Engineer of the North British Railway, the structure being 450ft. long, 250ft. wide and 78ft. 10in. high at maximum. The original station, of which some traces survive in the vicinity of Platform 1, was very limited in size; during the rebuilding, indeed, it was necessary to open out about 170 yards of the adjacent tunnel (which began at what is now the bridge carrying Cathedral Street over the railway, where the present glass roof ends) into cutting.

The city end of the station was extensively rebuilt during the later 1960s (when the terminus had to cope with traffic diverted from neighbouring Buchanan Street Station, closed in November 1966). The modernisation included a widened concourse, a new travel centre, and an extensive car park and taxi-rank on the site of the former goods station. Since then there have been further improvements, including the provision of a 'Casey Jones' fast-food restaurant and take-away (in 1983) and resurfacing of the concourse with a form of glazed white 'terrazzo' tiling.

Queen Street Station, Platform 6. ***ScotRail***

COWLAIRS TUNNEL Immediately beyond Queen Street Station is the entrance to the 1,000-yard Cowlairs Tunnel, excavated to carry the line below the short 'Cut of Junction' Canal which formed a connection between the Forth & Clyde and Monkland canals at the Townhead of Glasgow. The original intention had been to bridge the Cut of Junction and form a railway terminus a short distance to the north of the eventual location, but the Forth & Clyde Navigation Company objected to this course of action, forcing the railway company to burrow below the Canal instead. There are two ventilation shafts – one of which seems very close to the Queen Street end, because of the shortening of the tunnel at this end c.1880.

MILEPOST 1 MILE One mile from Queen Street, on the up side of the line and nestling between ballast and a retaining wall, is one of the original cast iron mileposts, installed when the railway was new in 1842. Most of the mileposts between Glasgow and Greenhill are still of this pattern, though none survive at other locations. Their design was inspired by turnpike road practice, and they still show mileages to Haymarket Station, Edinburgh, which was the eastern terminus for only the first four years of the line's history (being close to the up running line they are much more easily observed from down Edinburgh-Glasgow trains).

COWLAIRS BANK-HEAD The line between Queen Street and Cowlairs is very steep, much of it at 1 in 41, and until recent years this necessitated special arrangements to assist trains up the hill. From 1843 until 1908 (with two short breaks in the early years) trains were wound up by means of a huge stationary steam engine at the bank-head, which activated a continuous cable along the whole distance of the incline. Ascending trains were linked to the main cable at Queen Street by means of a 'messenger' chain, one end tied to the cable and the other hooked round an inverted draw-bar on the train engine buffer-beam. The 'messenger' chain was 'slipped' by the locomotive at the top of the incline, enabling the train to proceed without stopping. From 1908, when the cable was dispensed with, until the end of the steam era banking engines, fitted with slip couplings, were provided at the rear of trains. The winding-engine house, however, survived *in situ* at the top of the incline until about 1962.

COWLAIRS CHORD At the time of writing (1991) a new spur line to Springburn is under construction at Cowlairs, on the down side, to enable local trains to proceed between Queen Street and Cumbernauld without reversing at Eastfield, a short distance along the line.

SITE OF COWLAIRS WORKS
1 MILE 924YDS

A small industrial estate on the up side of the line (with premises including those of 'Bowmore Auchentoshan Glen Garioch Scotch Malt Whiskies') marks the site of the famous Cowlairs Works, which was built to serve as the principal railway works of the Edinburgh & Glasgow Company at the time of commencement of operations and survived in use until 1966. Locomotives were constructed between 1844 and 1924. Adjacent was a passenger station, intended both for the railway workers and the nearby Springburn community. It closed in 1964.

COWLAIRS WEST JUNCTION

Just beyond the site of Cowlairs Works a line branches away to the west, at a location formerly known as Helensburgh Junction (because it was the junction for the Glasgow, Dumbarton & Helensburgh Railway, opened in 1858). It subsequently became also the junction for West Highland Line trains to Fort William and Mallaig, when the West Highland route was laid from the Helensburgh line at Craigendoran. Nowadays trains to Oban also use this route.

EASTFIELD MOTIVE POWER DEPOT

The extensive railway establishment on the down side at this point began life as the Eastfield Engine Shed of the North British Railway, built in 1904 to replace a very cramped and overcrowded shed adjacent to Cowlairs Works. It was destroyed by fire in 1919, but was rebuilt and later modernised, in BR days, as a depot for diesel locomotives.

COWLAIRS SIGNAL BOX
1 MILE 1154YDS

Just beyond Cowlairs West Junction, on the up side, is Cowlairs Signal Box, which replaced eight old-fashioned mechanical boxes in the area when it opened in 1956. It later also took over the operations of the former mechanical signal box at Queen Street, on a separate panel on its operating floor. It was also able to absorb the bidirectional signalling scheme of the late 1980s between Cowlairs and Queen Street,

Eastfield Depot. ***ScotRail***

which was designed to increased line capacity on the incline and simplify working arrangements. Beyond Cowlairs Signal Box, again on the up side, is Cowlairs East Junction, which was laid in 1876 to provide a through route for goods and passenger trains from the east to the Helensburgh line.

BISHOPBRIGGS STATION
3 MILES 176YDS

The station at 'Bishops Bridge', as it was originally known, was one of the first E & G stations of 1842, though to begin with it was designated a 'temporary' station, until it could be ascertained whether enough traffic would materialise. It was scheduled for closure during the 1960s, under the Beeching Plan, but local campaigning bore fruit and it was reprieved. The substantial station building on the up side was, however, demolished and replaced by a small booking office and waiting shelter on the down side.

BISHOPBRIGGS OIL TERMINAL

At the end of a cutting beyond Bishopbriggs Station, on the up side, is Bishopbriggs Oil Terminal, opened in July 1971 by the Shell-Mex & BP Group. Spread over a 24-acre site, it cost about £1.5 million to build. Supplies of petroleum are brought by rail from a refinery at Grangemouth and distributed from here by road.

CADDER YARD
4 MILES 1154YDS

Immediately beyond the oil terminal is the site of Cadder Marshalling Yard, developed by the North British Railway during the early years of the century. With the modern changes of emphasis in British Rail freight it closed during the 1970s. Some of the lines on the down side remain in railway departmental use, but these on the up side have all been lifted and replaced by a large tree-planting project.

CADDER EAST SIGNAL BOX

Cadder East Signal Box, situated on the down side of the line near the east end of the former marshalling yard is one of the few traditional boxes on the route to survive.

LENZIE STATION
6 MILES 308YDS

The present Lenzie Station is on the site of the original 'Kirkintilloch' Station which was of a temporary nature. In 1844 the 'Kirkintilloch' station was moved a short distance eastwards along the line to a site at Garngaber, to enable an exchange platform to be established with a spur from the 4ft. 6in. gauge Monkland & Kirkintilloch Railway. When the new Campsie Branch was opened in 1848, diverging from the main line just east of the former station site, the station was moved back to its original location so that it could be served by the Glasgow-Lennoxtown trains which used the new branch. It was then renamed 'Kirkintilloch Junction', soon afterwards 'Campsie Junction', and in 1867 'Lenzie Junction' (later just 'Lenzie'). New, permanent buildings were established after the 1848 move, but these were taken down in 1977, to be replaced with the present modern facilities.

LENZIE 'TOWER'	A short distance north of Lenzie Station, on the brow of a hill, beyond a church, can be seen a large villa with a 'tower' feature. This building, dating from 1858, was one of the houses built under the 'villa tickets' scheme, which offered free travel to persons building large houses along the line during the 1850s.
CAMPSIE BRANCH JUNCTION	Immediately beyond Lenzie Station, on the up side, can be discerned the site of the former Campsie Branch junction, which was laid in 1848. Extending at first to Lennoxtown, via Kirkintilloch, the branch eventually reached as far as Aberfoyle. The Kirkintilloch-Aberfoyle section was closed to passengers in 1951 and the Lenzie-Kirkintilloch section likewise in 1964. Complete closure took place in 1966.
GARNGABER JUNCTION (SITE) 6 MILES 924YDS	In bushes about one-third of a mile beyond Lenzie Station, on the down side, can be noted the track of a former line curving away to the south. This was originally a link line to the ancient Monkland & Kirkintilloch Railway (which passed beneath the E & G line just east of here) and was built by the Edinburgh & Glasgow Company to the M & K 'narrow' gauge of 4ft. 6ins. Passengers travelling from Airdrie to Glasgow by this route required to change over at the 'exchange platform', mentioned above, to 'standard' (i.e. 4ft. $8^1/2$ in.) gauge trains on the E & G line. This situation prevailed until 1847 when the M & K gauge was altered to standard.
GARNGABER VIADUCT	Immediately beyond the site of the former exchange platform is a 5-arch viaduct carrying the E & G main line over the Bothlin Burn and the trackbed of the closed Monkland & Kirkintilloch Railway (opened 1826). For a short period, during the 1840s, a horse-drawn railway vehicle conveyed passengers from a platform below the viaduct to the town of Kirkintilloch. A flight of steps connected this platform with the E & G station above.
WOODILEE HOSPITAL SIDING (SITE) 7 MILES 594YDS	The Bothlin Burn, below the viaduct, runs through the grounds of Woodilee Mental Hospital, opened in 1875 by the Barony Parochial Board (Glasgow) to accommodate 'pauper lunatics'. The imposing twin-towers of the hospital were a notable landmark on the north side of the railway, until 2nd June 1990, when they were destroyed by fire. A siding was laid down at the hospital site as early as 1872, presumably at first for the purpose of bringing in building materials. It was removed in 1963, after a period of disuse.

WATERSIDE JUNCTION
8 MILES 418YDS

Beyond the hospital the railway enters a short, but fairly deep cutting. On emerging from this, the trackbed of a former branch line can be seen converging on the down side of the line. This was the double-track 'Waterside Loop', opened by the North British Railway in 1895, to provide a direct route from the east of Scotland to industrial Lanarkshire. It connected with the Monkland & Kirkintilloch line at Bridgend Junction, a mile to the south. The 'Waterside Loop' joined the main E & G line at Waterside Junction, where a signal box was opened, on the down side, at this time. The loop was closed and lifted during 1959-60; Waterside Junction signal box lasted a little longer, until 1962. Nearby the E & G line crosses over the Luggie Water by means of a small 3-arched viaduct.

WESTER GARTSHORE COLLIERY (SITE)

Immediately to the north a large colliery bing survives. This is Wester Gartshore, which was first rail-served in January 1889. The pit closed in March 1950, although the siding remained in position for some time afterwards.

GARTSHORE 9/11 COLLIERY (SITE)
9 MILES 616YDS

About a mile further on, traces of the former Gartshore 9/11 Colliery survive on the down side. To serve these workings 'Bairds New Junction Sidings' were opened on 12th September 1865 (although William Baird & Co. had laid down track within the colliery area as early as 1861). On the up side 'Bairds Siding' Signal Box (now demolished) was built to control entry to the colliery lines. Gartshore 9/11 Colliery closed in 1968. During its later years the famous Neilson 0-4-0 saddle tank of 1876, 'Kelton Fell' (latterly 'No.13' in NCB ownership), could often be seen shunting here. It is now preserved by the Scottish Railway Preservation Society at Bo'ness.

CROY STATION
11 MILES 616YDS

Croy Station, almost 11½ miles out from Glasgow, was originally constructed as a 'temporary' station. Closure notices were posted on 10th December 1965 – but the station 'weathered the storm' and has survived. A typical NBR station building is still in existence on the down platform, although largely boarded up to prevent vandalism. A particularly severe attack of vandalism at the station was reported in the Glasgow *Evening Times* of 26th March 1988.

CROY CUTTING
12 MILES

Just beyond Croy Station is a formidable rock cutting, in places 70ft. deep. Its excavation caused the Edinburgh & Glasgow Railway a great deal of concern, and the work necessary here delayed the opening of the line by some months. The contractor used about 25 tons of gunpowder in blasting a way through.

Local train approaching Croy cutting from the west. ***ScotRail***

DULLATUR STATION (SITE) 12 MILES 1254YDS

Not far beyond Croy cutting was Dullatur Station, of which little trace survives. A clue to its position is given, however, by a terrace of substantial stone-built residential houses on the down side. The station was directly opposite. It was not one of the original E & G stations and in fact did not come into use until 1876, thirty-four years after the line's opening. Closure notices were posted on 10th December 1965, the same day as those for Croy, but unlike Croy, it did not weather the storm, and closed with effect from 5th June 1967.

CASTLECARY STATION (SITE) 15 MILES 682YDS

About 2 miles beyond the Dullatur Station site, at the end of a long, shallow cutting, is the site of another former station, known as Castlecary. This was one of the original stations of 1842, and indeed was the first one out from Glasgow to be designated 'permanent' at that time, and stone buildings were provided. Although serving a very small community the station lay beside an important road, and a connecting coach service was established between here and Stirling. A timetable of 1844 shows all trains stopping here, the only two other locations as well provided being Falkirk and Linlithgow. The importance of the road connection declined after the opening of the Scottish Central Railway to Stirling in 1848 and the station's purpose for most of its life was merely to serve a small village. It succumbed during the Beeching cuts of the 1960s, closing to passengers with effect from 6th March 1967. The adjacent signal box had closed a little earlier, on 6th November 1966.

CASTLECARY VIADUCT

Immediately beyond the Castlecary Station site the line runs onto Castlecary (or Redburn) Viaduct, built over the main Glasgow-Stirling Road (now the M80). The 8-arch viaduct is 200 yards long and has a maximum height of 150ft. over the valley below. Constant use over 150 years has taken its toll, however, and some of the arches now bear bracing to maintain the integrity of the structure.

CALEDONIAN RAILWAY BRIDGE

A short distance beyond Castlecary Viaduct the E & G line crosses another double-track main line railway. This was opened on 7th August 1848 as the Castlecary Branch of the Caledonian Railway, part of a link between the principal CR Glasgow-Carlisle route and the Scottish Central Railway at Greenhill (Lower). In effect, it provided a direct route from England, via Carlisle, to the northern parts of Scotland, bypassing Glasgow and Edinburgh. It was extensively employed as such by, successively, the Caledonian Railway, the LMS Railway and British Railways, although to a much lesser extent in recent years. No junction with the Edinburgh & Glasgow line exists at this point, although at one time it was planned to form one, and construction of a fairly substantial embankment was begun to the south of the E & G. Its incomplete remains can still be discerned, although heavily overgrown.

Castlecary Viaduct. ***ScotRail***

GREENHILL UPPER JUNCTION
17 MILES 198YDS

About a mile and a half beyond the Caledonian bridge, at Greenhill Upper Junction, a double-track line forks away to the north-east, on the up side of the E & G. This line dates from 1848 and was opened on 22nd May of that year by the Scottish Central Railway, as part of a trunk route which stretched from here northwards to Perth, via Stirling. An anomaly was created in 1865 when the Scottish Central Company was absorbed by the Caledonian Railway, at about the same time as the Edinburgh & Glasgow Railway Company became part of the rival North British Railway Company. Trains from Perth, Stirling and other northern parts continued to run by way of Greenhill Upper into Glasgow (Queen Street) Station for a while, however, until diverted by the Caledonian into Glasgow (Buchanan Street) via their own route from Greenhill Lower in 1870. In recent years the Queen Street-Greenhill Upper route has been restored as the principal main line from Glasgow to north-central and north-east Scotland.

Greenhill to Polmont via Falkirk High

BONNYBRIDGE HIGH STATION (SITE)
18 MILES 770YDS

Less than a mile beyond Greenhill Upper Junction a cluster of houses beside an underbridge marks the site of Bonnybridge High Station. The village proper lies about a mile to the northwest, but the station was nevertheless known simply as 'Bonnybridge' when opened by the North British Railway. The 'High' epithet was added in early BR days, by which time, ironically, the two other stations serving Bonnybridge, on different lines, had been closed to passengers. Bonnybridge High was itself closed as a passenger station in March 1967, by which date the adjacent signal box (situated to the east of the station, on the up side) had also been closed.

SUMMERFORD BRIDGE

About two miles past the site of Bonnybridge High Station, beyond the second of two short cuttings, the railway crosses a high bridge, to the north of which a public road leads at right angles. This is Summerford Bridge, which was built to supersede a viaduct of the same name in 1961. Summerford Viaduct had six arches, one of which was considerably larger than any of the others, with a span of 130ft. Through it passed a lock-flight of the Union Canal, and a branch railway serving numerous industrial concerns to the north. Although the full width of the arch was 130ft. it was only 24ft. 6in. in height, and was claimed to be the largest span in Britain with such a low elevation. The roadway passed through one of the smaller arches, on an alignment some distance to the west of the present road. The new bridge of 1961 was built a short distance to the south of the old viaduct, requiring new embankments to be created for the rail approaches. That for the eastern approach was constructed on top of the trackbed of the former branch railway, which had passed through the main arch of the old viaduct, then curved to the east to run parallel with the E & G main line before joining it near Falkirk High Station. The canal lock-flight also turned to the east after passing through the viaduct arch. The remains of one of the lock chambers can be seen from the train, within a picnic site on the south side of the line, a short distance further on.

FALKIRK HIGH STATION
21 MILES 1144YDS

Although not very convenient for the town centre, Falkirk High is one of the principal stations on the Edinburgh & Glasgow line and has been so since the opening of the railway in 1842. Substantial stone buildings were originally provided on the up platform, but in 1980 these were demolished and superseded, on the same site, by a neat structure of rustic facing brick, incorporating a booking hall/ ticket office, heated waiting room, toilets and staff accommodation. A heated waiting shelter was at the same time provided on the down platform. The replacement buildings (which cost £120,000) were inaugurated on 8th December 1980 at a ceremony performed by James Anderson, Convener of Central Regional Council, a locomotive driver by profession. Councillor Anderson then drove the 09.55

Falkirk High Station, with Class 158 diesel train approaching the 'up' platform. *Strathkelvin District Libraries*

train to Glasgow! The goods yard at Falkirk High, on the up side of the line, was closed and all sidings lifted in 1964, after which the ground was converted into a car park. The adjacent signal box closed on 4th February 1968.

FALKIRK TUNNEL

Immediately to the east of Falkirk High Station is Falkirk or Callendar Tunnel, 845 yards long. It was built for the opening of the line, and immediately before coming into use it was illuminated internally and thrown open for public inspection, at a small charge per head in aid of charity. Over the years it came to be regarded as an operating liability, because of inadequate drainage, limited clearances and necessary speed restriction. Following the Penmanshiel Tunnel collapse (on the Edinburgh-Berwick line) of 1979, a careful examination of Callendar Tunnel was made. In the light of this it was decided to close it for extensive repairs and consolidation. The decayed condition of old mineworkings in the vicinity was considered to be a particular hazard. Infilling with grout and other engineering works, including rock bolting at the Glasgow end and replacement of old brick lining by reinforced concrete sections at the eastern end, was carried out at a total cost of about £2.5 m. For much of the year 1980 all trains were diverted via Falkirk Grahamston to allow work on the tunnel to be carried out. Reopening in December coincided with the opening of the rebuilt Falkirk High Station, and also with the introduction of a full service of push-pull trains of Mark 3 carriages, operated by Class 47/7 locomotives.

POLMONT ENGINE SHEDS (SITE)

Two miles beyond the eastern portal of Falkirk tunnel, a siding on the down side of line marks the site of the now-vanished Polmont Engine Sheds, built by the North British Railway in 1913-15 to supersede a previous facility at Manuel (Bo'ness Junction), between Polmont and Linlithgow. During the early 1920s the allocation was 44 locomotives, mostly for working mineral traffic originating in the Falkirk-Polmont-Bo'ness area. During the late 1950s and early 1960s the shed yard was notable for the number of interesting locomotive types to be seen in store, including examples of different 4-4-0 classes. At this period, the working activities of the depot were in irreversable decline, however, and it closed completely on 17th May 1964, when its work was transferred to the former Caledonian Railway shed at Grangemouth.

Immediately beyond the Polmont Shed site, the former Stirlingshire Midland Junction line, from Falkirk Grahamston, converges on the up side, at Polmont Junction.

Polmont to Edinburgh (Waverley)

POLMONT STATION
24 MILES 1408YDS

Polmont was one of the 'permanent' stations provided by the E & G Railway Company for the line's opening in 1842. The original main building survives on the up side, extensively modernised during the mid-1980s as part of a £250,000 refurbishment of the station. A commemorative plaque was unveiled on 25th July 1985 by Councillor James Anderson, Convener of Central Regional Council, to mark the completion of the project. Another important ceremony took place at Polmont Station on 23rd July 1990 when Ian Lang, then Minister of State at the Scottish Office, launched the 'Green Link' project, to improve the Edinburgh-Glasgow railway environment, by planting a tree there. Notices giving information about related canal/railway walks have been erected at Polmont, Falkirk High and Linlithgow stations. The site of a former bay platform for Bo'ness branch passenger trains, on the up side, is now occupied by grass and shrubs.

BO'NESS (HIGH)
JUNCTION
26 MILES 1672YDS

Trains to Bo'ness from Polmont formerly left the main line at Bo'ness (High) Junction, just over two miles from Polmont. The northward curving branch to the Forth port of Bo'ness was first opened in 1851 (to passengers in 1856) but lost its passenger service in May 1956. Complete closure of the

Polmont Station, showing refurbished buildings on the 'up' platform. ***ScotRail***

branch took place in 1980, but in recent years it has been progressively re-opened, from the Bo'ness end, by the Scottish Railway Preservation Society, culminating in 1990 with a re-connection to the main line. The direct connection which had previously existed from the Polmont direction was, however, replaced by a 'Y' formation, making a trailing connection into the up line some distance to the east.

MANUEL STATION (SITE)
27 MILES 836YDS

The new connection has, ironically, removed the final traces of Manuel passenger station, which was on two levels, with a secondary line passing below the main line at right angles. The lower line was built in 1851 as the Monkland Railways' direct line from Lanarkshire to Bo'ness. Connection with the line from Bo'ness (High) Junction was made a short distance to the north, at Bo'ness (Low) Junction, and a passenger service was established in 1856. Low level passenger trains from the Lanarkshire direction ceased as from 1st May 1930, although an infrequent shuttle from Manuel (Low Level) to Bo'ness lingered for a short while thereafter. Manuel (High Level) Station, on the main E & G line, closed in March 1967. Right to the end booking-office facilities were provided on the lower level. Although all traces of the station have now gone, parts of the track alignment of the low level line can still be made out, both north and south of the main line.

AVON VIADUCT

The western approach to Linlithgow is over the spectacular 23-arch Avon Viaduct, now Listed Grade A and scheduled as an Ancient Monument. It was completed in 1841. The railway company then received an unpleasant shock when the Royal Burgh of Linlithgow claimed the right, under mediaeval charter, to levy 'customs duty' on traffic crossing the River Avon over the viaduct. The Burgh already enjoyed this privilege with regard to the nearby road bridge and stood to lose revenue with the coming of the railway. Litigation was prolonged and expensive, and it was not until 1853 that the House of Lords finally found in the railway company's favour. The present Linlithgow Bridge, three hundred yards to the north of the railway viaduct, replaced a seventeenth century bridge in 1960.

LINLITHGOW STATION
29 MILES 946YDS

Linlithgow has one of the original E & G stations, dating from 1842. A bright new facility has in recent years been created 'within the fabric' of the original main station building, on the up side, at a cost of £315,000. It incorporates a modern ticket office and staff accommodation at ground level, with waiting room and toilet facilities at platform level above. The modernised station was officially opened on 4th November 1985 by Tam Dalyell, the well-known local M.P. At the same time locomotive No.37401 was named *Mary*

Avon Viaduct, Linlithgow, viewed from the west. ***ScotRail***

Linlithgow Station. Refurbished buildings on the 'up' platform. ***ScotRail***

Queen of Scots in honour of the town's best-known native. An original stone-lined underpass between the platforms leads directly to a modern waiting room on the down side.

PHILPSTOUN STATION (SITE)
32 MILES 1364YDS

About three miles beyond Linlithgow the small village of Philpstoun comes into view on the down side, close to the railway. The station was opened in October 1885 and retained a passenger service until June 1951, when it was withdrawn. The buildings were demolished during 1955.

WINCHBURGH JUNCTION
34 MILES 946YDS

Immediately beyond Philpstoun the railway enters a deep rock cutting, which leads to Winchburgh Junction. The Junction was created in 1890 when a new line was laid, diverging on the north side, to provide a direct connection from the west towards the newly-built Forth Bridge. The dominantly-positioned signal box (a standard North British 'Brick and Timber' structure of the period), high on the north side of the cutting, was a well-known railway landmark until its closure in February 1979. The line from here to Dalmeny and the Forth Bridge has not been used by regular passenger trains since 1973 but is often brought into use as an emergency option. From time to time a proposal has surfaced to divert the principal Glasgow-Edinburgh trains by this route to near Dalmeny, and from there onwards by the Forth Bridge-Edinburgh line. To accomplish this in an acceptable way a new curve would require to be laid south of Dalmeny, to facilitate through running without reversal. The advantage of this scheme would be to eliminate expensive maintenance work on Winchburgh Tunnel and Almond Viaduct.

WINCHBURGH STATION (SITE)
35 MILES 462YDS

A short distance beyond Winchburgh Junction, in deep cutting, is the site of the former Winchburgh Station. Much of Winchburgh village lies above Winchburgh Tunnel, just beyond, and a footpath led back along the east side of the cutting, at high level, to the station buildings, which were also located in an elevated position (apart from waiting rooms, at platform level). Winchburgh was one of the original E & G stations of 1842, but closed early, in September 1930. Traces of the stairways which led down to platform level can still be easily discerned.

WINCHBURGH TUNNEL

Just beyond the Winchburgh Station site is the 372-yard Winchburgh Tunnel. Clearances within the tunnel are very tight, it is susceptible to flooding, and for long it was subject to a speed restriction of 60 miles per hour (now relaxed to 80 miles per hour).

Almond Viaduct, viewed from the west. ***ScotRail***

ALMOND VIADUCT

Preceded by a short 7-arch viaduct over the A89 Newbridge-Broxburn road, approach is made to the most spectacular feature of the Edinburgh & Glasgow Railway, the half-mile long Almond Viaduct. Construction commenced in October 1838 and occupied a period of some thirty-one months. The viaduct consists of thirty-six arches of 50ft. span and 70ft. height and contains an estimated 1,105,000 cubic feet of solid masonry. It cost approximately £130,000 at the time it was built. Age has since taken its toll of a structure which now carries loads of a weight and at speeds unheard of when it was designed, and cracking has appeared, but not even the rather unsightly bracing of old bullhead rails which supports the arches can detract from the magnificent outline of the viaduct, built on a curve. It is a listed building, Grade A.

M8 BRIDGE

Beyond the Almond Viaduct the railway crosses the M8 Glasgow – Edinburgh motorway on a modern steel girder bridge, with concrete deck, commissioned in July 1970. It has a 113 ft span and weighs approximately 500 tons.

BATHGATE JUNCTION
38 MILES 1078YDS

The 10-mile line from the south side of the E & G line at Bathgate Junction to Bathgate was opened by the Edinburgh & Glasgow Railway on 1st May 1850. With the completion of a line between Bathgate and Airdrie by the Monkland Railways during the early 1860s and later improvements in the Glasgow area the Edinburgh & Bathgate line became part of an alternative through route between Edinburgh & Glasgow, although this was closed to passengers between Bathgate Junction and Airdrie in January 1956. At a later date most of the line between Bathgate and Airdrie was closed to all traffic and lifted. In recent years there has been a remarkable revival, for on 24th March 1986 the section between Bathgate Junction and Bathgate was reopened to passengers and a regular service established between Edinburgh Waverley and Bathgate, also serving intermediate stations on the branch at Uphall and Livingston North. Notorious road congestion on the western approach to Edinburgh has ensured the popularity of this reinstated service. Bathgate Junction is nowadays known as Newbridge Junction.

RATHO STATION (SITE)
38 MILES 1694YDS

Ratho was one of the first 'permanent' stations when the E & G line was opened in 1842, even though it served a comparatively small village. Its passenger service was withdrawn in June 1951, although platforms and buildings were maintained in good order for some considerable time afterwards. They have deteriorated in recent years, but the north platform surface can still be easily identified from passing trains.

GOGAR STATION (SITE)
41 MILES 1144YDS

Gogar was not one of the original E & G stations, but was nevertheless opened very early in the line's history. It lost its passenger service in September 1930. Following a period of dereliction, the westbound platform was removed in 1959, the goods yard was closed in 1960, and finally the empty station buildings (on the north side of the line) were demolished during the early 1960s. Very little trace of the station now survives.

SAUGHTON JUNCTION
43 MILES 1100YDS

Saughton Junction was created in 1891-92 when a new line was laid, diverging on the north side of the E & G line, to provide a direct connection from the Edinburgh direction towards the newly-built Forth Bridge The junction was situated immediately east of the existing 'Corstorphine' passenger station, one of the original E & G stations of 1842 (though initially designated a 'temporary' station). The station house was on the north side of the line. After the opening of the Forth Bridge line an extra pair of platforms were provided to serve this route. However, the station was some considerable distance from the village it purported to serve and the use of it was greatly affected when the NBR opened a new branch from Haymarket (West) Junction to central Corstorphine on 1st February 1902. The E & G station was then named 'Corstorphine (Old)' and later renamed 'Saughton Junction'. It was closed to passengers in 1921. The line between here and Haymarket Station was quadrupled in 1891-92. Trains from the Forth Bridge direction cannot actually pass on to the E & G line here, there being, in effect, an 'up' and 'down' line for each route from here into Edinburgh.

CORSTORPHINE BRANCH JUNCTION (SITE)

Some distance further along the trackbed of the former Corstorphine Branch will be seen, converging on the left-hand side, on embankment. As already mentioned it opened on 1st February 1902. A once-familiar sight, right beside the junction, although serving branch trains only, was the station of 'Balgreen Halt', opened by the London & North Eastern Railway in 1934. The branch was not scheduled for closure by Dr Beeching in his report on *The Reshaping of British Railways* (1963), but was closed just the same – to passengers in December 1967, and to goods in February 1968.

HAYMARKET WEST JUNCTION 44 MILES 1518YDS

As well as controlling entry to the Corstorphine branch, the former Haymarket West Junction signal box controlled entry to two other lines, both on the down side, here. The first was a link-line to the former North British Edinburgh suburban loop, which formed a diversionary route to take traffic away from the busy Edinburgh Waverley route. The second line (now closed), linked with the former Caledonian Railway route into Princes Street Station, so enabling Caledonian trains from the Stirling direction to use the Stirlingshire Midland Junction and E & G routes *en route* to Princes Street.

At Haymarket Central Junction, further on, trains from the Suburban loop formerly gained access to the main line, for entry into central Edinburgh. This route survives as a single line of track.

HAYMARKET SPRINTER DEPOT

On the up side, immediately beyond Murrayfield Stadium, is the large motive power depot of Haymarket. It was built in 1892-93 to replace a much more congested site nearer Haymarket Station. For most of its history it housed and serviced steam locomotives, but was closed to steam in September 1963 and reconstructed to form a diesel maintenance depot. More recently, it has been further modernised as a depot for diesel units of classes 150, 156 and 158. This 'Haymarket Sprinter Depot' was officially opened on 12th October 1988.

Haymarket Sprinter Depot, with Class 158 diesel train on Edinburgh-Glasgow service in foreground. ***ScotRail***

Haymarket Station. Original building at street level, after stone cleaning. ***D. Martin.***

New footbridge in Princes Street Gardens. ***ScotRail***

HAYMARKET EAST JUNCTION

Close to Haymarket Station, on the down side, a line bearing electric wires converges, at Haymarket East Junction. This is the so-called 'Duff Street Connection' opened on 7th September 1964 to enable Princes Street Station to be closed and trains from the former Caledonian line to enter Haymarket and Waverley stations instead. It has recently been electrified, to allow West Coast Main Line trains direct entry to Waverley 'under the wires'.

HAYMARKET STATION
45 MILES 1716YDS

The original Haymarket Station, with a train shed alongside the present station site, to the north, came into use when the E & G railway opened in 1842. It was then the terminus of the line. 'Through' platforms came into use after 1846 when the railway was extended to Waverley. There have been further extensions and modifications to the station, extending over a long period. During the early 1980s extensive modernisation took place, the refurbished station being opened by Lord Provost Tom Morgan of Edinburgh on 23rd March 1984. It was claimed to be 'the first purpose designed "open" station to be completed in Scotland' (i.e. with no ticket barriers). The original train shed, latterly redundant, was removed during 1982-83 to the Scottish Railway Preservation Society station at Bo'ness. However, the original entrance building at street level above was stone-cleaned, refurbished and retained for ticket-office and other accommodation.

HAYMARKET TUNNELS

There are two double-track tunnels between Haymarket Station and Princes Street Gardens, side by side. The first was opened with the original railway extension in 1846, the second, to the south, was excavated during the early 1890s, being opened officially on 21st July 1895. The tunnel portals at the east end of Haymarket Station have a distinctively different appearance, the north portal being at right angles, with the southern one skewed.

PRINCES STREET GARDENS

Beyond the tunnels, the line opens out into the pleasant Princes Street Gardens, with the castle towering above on the right-hand side. The entry of the railway into the Gardens was furiously opposed, and has been a matter of controversy ever since. More information concerning the railway through the Gardens will be found in the 'History from the Train' section.

WAVERLEY STATION
47 MILES 242YDS

At the eastern end of the Gardens the railway passes through The Mound, bearing the Royal Scottish Academy and the Scottish National Gallery, and enters Waverley Station. The station has been rebuilt (and indeed renamed) on a number of occasions, notably in 1894-1900, when the station was extensively remodelled. The extra traffic generated by the opening of the Forth Bridge had resulted in chronic congestion, bordering on chaos. Developments in recent years have included a new Travel Centre, opened by the British Rail Chairman, Sir Robert Reid, on 16th May 1986. In recent years there has been further talk of redevelopment, perhaps with a 'roof' over the main station area to give Edinburgh some additional prime development space. Waverley Station is now 'listed', Grade A.

Waverley Station. Travel Centre and waiting area. ***ScotRail***

Greenhill to Polmont via Falkirk Grahamston

Some Edinburgh & Glasgow line services diverge north-eastwards from the main route at Greenhill Upper Junction, and follow an alternative course to Falkirk Grahamston Station and/or beyond, notably local trains between Glasgow Queen Street and Grahamston, serving Bishopbriggs, Lenzie and Croy.

GREENHILL LOWER JUNCTION
17 MILES 1562YDS (FROM GLASGOW)

From Greenhill Upper the route followed is that of the Scottish Central Railway, opened in May 1848 from here to Perth. A short distance along, at Greenhill Lower Junction, a double track line converges on the up side. This is the former 'Castlecary Branch' of the Caledonian Railway, opened in August 1848 as part of that railway's connecting line between their main Carlisle-Glasgow route and the Scottish Central line to the north. It is still available today as a through route between England and the north-east of Scotland, avoiding Glasgow and Edinburgh. Adjacent to the junction was the Caledonian station of Greenhill, closed to passengers on 18th April 1966.

BONNYBRIDGE BRANCH (SITE)

Just beyond Greenhill Lower Junction can be seen traces of the trackbed of the former Caledonian branch to Bonnybridge (Canal), opened in 1886. It seems to have been built at the instigation of industrial customers of the Caledonian Railway locally, who had previously been served by the Forth & Clyde Canal (also Caledonian-owned) for their transport needs. They felt that they were placed at a disadvantage to rivals served by branches of the North British Railway, and so obliged the Caledonian Company to build a branch railway to compete with its own canal. It seems that the Bonnybridge Branch was known locally as the 'Penny Line', perhaps a reflection of the fare between Bonnybridge and Greenhill. The branch was closed to passengers in July 1930, and to goods in December 1964.

FORTH & CLYDE CANAL AQUEDUCT

About two miles beyond Greenhill Lower Junction, trains between Glasgow and Falkirk Grahamston still require to pass below an aqueduct carrying the Forth & Clyde Canal. The canal at this point was opened for traffic in 1773, and it was over seventy years later that Scottish Central Railway engineers were faced with the formidable task of burrowing below its calm waters.

CARMUIRS WEST JUNCTION
20 MILES 550YDS

A short distance beyond the canal aqueduct is Carmuirs West Junction, where Glasgow-Grahamston trains diverge from the Scottish Central Railway route of 1848 onto a short spur line (from here to Carmuirs East Junction) built under a Caledonian Railway Act of 1976, and opened in 1882, to facilitate through running between Glasgow (Buchanan Street) and Grangemouth.

CARMUIRS EAST JUNCTION
20 MILES 1386YDS

At Carmuirs East Junction the 1882 spur line joins the Stirlingshire Midland Junction line from the Scottish Central at Larbert Junction, to the north, to Polmont via Falkirk Grahamston. It was opened in 1850, by which date it had already been absorbed by the Edinburgh & Glasgow Railway Company. Construction of the junction at Carmuirs East was undertaken by the North British Railway at Caledonian Railway expense. The signalbox at Carmuirs East was erected to North British designs and signalling between Carmuirs Wast and Carmuirs East was under the North British Block Telegraph regulations, with NBR instruments.

CAMELON STATION (SITE)

Just over half a mile beyond Carmuirs East Junction a separation of the tracks marks the site of the former island-platform station of Falkirk Camelon, closed to passengers in 1967.

CANAL SWING BRIDGE
21 MILES 1650YDS

Half a mile beyond the site of Camelon Station the railway crosses the Forth & Clyde Canal by a 'swing' bridge (which no longer swings). The present structure replaced an earlier swing bridge in 1899, but of course has not required to be operated since closure of the canal at the end of 1962. Just beyond the bridge is 'Swing Bridge East' signalbox, still operational. There was previously a 'Swing Bridge West' box on the other side of the canal, required to control a former level crossing nearby. In 1927, however, a cable was laid across the bed of the canal to enable Swing Bridge East box to take care of the crossings.

FALKIRK GRAHAMSTON STATION
22 MILES 902YDS

Another half-mile onwards is the important station of Falkirk Grahamston, dating from 1850. This station is much more convenient for Falkirk town centre than Falkirk High Station on the Edinburgh & Glasgow main line. The original Stirlingshire Midland Junction station building of 1850, on the down side , was demolished during 1985-86, along with other old buildings on both platforms. It was replaced with a smart new building on the up platform, incorporating ticket office, waiting hall, public toilets, bookstall, parcels point and offices. An enclosed waiting shelter was at the same time provided on the down platform. The official opening of the new station buildings was carried out by Councillor

Falkirk Grahamston Station. ***ScotRail***

Swing Bridge, over the Forth & Clyde Canal, at Falkirk. ***Strathkelvin District Libraries***

James Anderson, Convener of Central Regional Council, on 28th April 1986. At the same time Class 47 locomotive No. 47636 was named *Sir John de Graeme* in honour of a Scottish patriot killed at the Battle of Falkirk in 1298. Falkirk Grahamston Signalbox was destroyed by fire in 1972.

GRANGEMOUTH JUNCTION

The 2-mile Grangemouth branch was promoted by the Forth & Clyde Navigation Company (i.e. the Forth & Clyde Canal) and was opened from the Stirlingshire Midland Junction line at Grangemouth Junction in 1860 (goods only); passenger trains commenced the following year. At first the trains on the branch were worked by the Edinburgh & Glasgow Railway Company, then from 1865 by the NBR, but in 1867 the Canal was purchased by the Caledonian Railway Company, who thus acquired the Grangemouth Branch Railway as well. Passenger services on the branch ceased in January 1968. Grangemouth Junction is situated about half-a-mile east of Grahamston Station. Nowadays the branch is used mainly by oil trains operated by Railfreight's petroleum sub-sector.

Down the Years

In the Beginning

Nowadays, when we complain of the inadequacy of some motorways to cope with the volume of traffic, or of persistently late-running trains on certain railway lines, it is easy for us to forget just how bad inland transportation really was until quite modern times. Today we are pampered and spoiled by comparison. In Scotland, for example wheeled vehicles of any kind were rare until about 250 years ago. They had existed in Roman times, but had disappeared again, not to re-emerge for over a millennium. Throughout the mediaeval period roads were quite awful, and seaboard towns placed at great advantage because they enjoyed, in coastal navigation, by far the most efficient form of transport then available. The roads were reduced to quagmires by driven cattle, and in many areas it was necessary to establish and maintain a narrow raised causeway on one side, for the use of pedestrians, packhorses and riders on horseback. Collectively, the packhorses formed a vital economic lifeline. During the eighteenth century turnpike trusts were formed to improve road surfaces at the expense of users, who paid tolls at turnpikes or toll-gates, and about the same time a few canals were constructed to extend the acknowledged benefits of shipload conveyance inland. Primitive railways also began to emerge, but it was a long time before they adopted anything remotely resembling their modern format. The rails were of wood or brittle cast-iron and horse power the only means of haulage. Public traffic was never carried, the lines being laid on private land for private use.

The turn of the nineteenth century ushered in the era of the *public* railway. In Scotland it began with the Kilmarnock & Troon, opened in 1811 – a remarkably enterprising and innovative railway, which in addition to carrying public traffic introduced a passenger service and brought the latest in technology to Scotland, in the form of steam locomotive power. Unfortunately for the Troon railway, technology in rail manufacture had not yet come to terms with weighty metal locomotives, so its primitive flanged plateway was badly knocked about by the 'new toy' from England, and the latter had to be laid aside. In 1826 the Monkland & Kirkintilloch became the first public railway in Scotland to be laid with Birkinshaw's patent malleable iron rails – the rails which revolutionised railway development on account of their ability to properly support steam locomotives and allow them to pass along at reasonably high speeds.

Around this time serious consideration was being given to the possibility a railway between Edinburgh and Glasgow. An engineer named James Jardine published a brief report for a line via Fauldhouse, Shotts and Carmyle (*Preliminary report to the Committee of subscribers to the survey of the Edinburgh, Leith, Glasgow, and Paisley Railway*) in 1825, and a prospectus, plans and estimates the following year. Nothing further happened at this time, but in 1830 Thomas Grainger and John Miller, erstwhile engineers of the Monkland & Kirkintilloch Railway but by this time in cahoots with the rival Garnkirk & Glasgow Railway, began to look at schemes for extending the short G & G line eastwards to Edinburgh. To create a favourable climate the help of the great George Stephenson was enlisted.

Fatherly Guidance

George Stephenson deserves the title 'Father of Railways' often accorded him. He was a colossus among men who strode through the vital early period of public railway development with the supreme confidence of an ultimate expert in the most significant technology of its time. Grainger and Miller must have been gratified that he was so ready to endorse their scheme, albeit with some modification. This intervention by Stephenson in Scottish railway affairs brought with it the benefits of current philosophy on the aims and purposes of railway promotion. In a report, dated 30th December 1830 'on Line of Edinburgh, Glasgow and Leith Railway', Stephenson described 'the expeditious, safe and cheap conveyance of passengers' as 'the object which, in all undertakings of this description, should be especially aimed at'. This ideal had been brought to fruition in England during 1830, with the opening of the Liverpool & Manchester Railway, but it was novel in Scotland, where all the lines then in existence had been built for the almost exclusive purpose of carrying coal. Stephenson gave consideration to James Jardine's 'Old' or 'South' line, and while recognising that it passed through a rich coal and iron field he took the view that Grainger & Millers' 'New' or 'North' line was more suited to the purpose of conveying passengers between the cities, being less hilly.

In a later report, dated 6th June 1831, Stephenson considered the possibility of a route by Falkirk, acknowledging that 'the valley along which the Forth and Clyde Canal passes rises to a much less height than any other valley running through the country from East to West', but he pronounced that 'owing to the extremely rough and hilly surface of the intervening country, the entrance into it, both from Glasgow and Edinburgh, could be effected only with an immense sacrifice of directness to the extent probably of ten miles, and by a ruinous expenditure of money in the formation of the line...' He therefore endorsed Grainger & Millers' scheme once again, which in its final form was intended to pass through Bathgate and Slamannan on its way from Edinburgh, before connecting with the Garnkirk & Glasgow line near Coatbridge. The supporters of the project were by this date able to point to the success of the Liverpool & Manchester Railway, during its first few months, and draw confident parallels with future Edinburgh-Glasgow traffic. Unfortunately their Bill failed at Commons Committee stage in May 1832, probably on account of opposition by local canal interests, who were at pains to demonstrate the *dissimilarity* between the Liverpool-Manchester and Edinburgh-Glasgow schemes.

Getting the Act Together

Despite the 1832 failure, it was really only a matter of time before a successful Edinburgh-Glasgow project was certain to emerge. Promotion of the favourably destined scheme began in 1835. Its intended purpose as an inter-city route for passengers and general goods was reflected in the predominance of city merchants among the promoters (the boards of the early mineral railways having been dominated by local industrialists).

Well over half the invested capital came from England, however, with Lancashire money being especially prominent – Liverpool-based subscriptions amounted to more than the combined total of those from Edinburgh and Glasgow added together. Control remained in Scotland, nevertheless, with only two Lancashire directors representing English money on the Board. The chairman, John Leadbetter, was a prominent Glasgow man, while his depute, John Learmonth, was a leading Edinburgh merchant. Both occupied senior positions on the boards of other leading Scottish railways, at different times.

The line adopted was the northern 'Valley along which the Forth and Clyde Canal passes' rejected by George Stephenson partly on account of the 'ruinous expenditure of money' estimated to be necessary. On this occasion it was decided to outlay the capital in order to achieve a level 'high-speed' route, to the north of the main mineral field. Its advantage over the 1830-2 route, in terms of ease of gradient, was proportionately much greater than that which the latter route enjoyed over James Jardine's proposed line via Shotts. The cost was estimated at £550,000, as compared with £410,000 for the 1830-32 scheme, as revised. Ironically, the new scheme could be described as 'Stephensonian' in concept – i.e. it followed Stephenson's ideals of bridging valleys and penetrating hills, rather than placing strain on his progeny locomotives, with too many gradients.

The earlier campaigns of 1830-32 caused the Edinburgh & Glasgow promoters of 1835 to take potential opposition from the canal companies very seriously indeed. The canal companies probably considered the eventual success of an Edinburgh-Glasgow railway scheme as inevitable, especially as during the early 1830s they had introduced a new fleet of fast, sleek narrow (though still horse-drawn) passenger boats, known as 'Swifts', in readiness for the anticipated competition, but they were not prepared to give up without a fight. The new Edinburgh & Glasgow promoters tried to play down the concept of railway/canal competition, targeting instead the existing road traffic.

One of the fast-moving horse-drawn 'Swift' boats introduced on the Forth & Clyde Canal during the 1830s, with a view to meeting the imminent rail competition. *Mitchell Library*

They proffered the view that minerals and other bulk goods would very likely stay with the canals, and indeed no mention of coal traffic was made in the E & G prospectus. The canal companies were not fooled, however, and embarked on a campaign of vigorous opposition. Their awareness of the railway threat is perhaps also reflected in the way that the Union Canal, which extended from Falkirk to Edinburgh, interested itself in railway schemes to link the Falkirk area with Glasgow, thus providing potential competition for Forth & Clyde Canal traffic between these places. Firstly, there was a scheme to connect Glasgow by rail with the vicinity of the Forth & Clyde/Union Canal junction at Lock 16, near Falkirk. In 1835 this was being promoted as the 'Edinburgh and Glasgow Junction Railway', although by the following year it was more accurately known as the 'Glasgow and Falkirk Junction Railway'. Little more was heard of this proposal, but the Union canal soon became deeply involved in the Slamannan Railway project to provide a railway/canal route between Glasgow and Edinburgh, with transfer of goods and passengers between modes at Causewayend, near Linlithgow. The Union Canal Company paid £500 towards the expense of obtaining the Slamannan Railway Act, and the cost of providing a basin and wharves at Causewayend was shared between the Canal Company and the Slamannan Railway Company. Traffic on this short-lived inter-city route began in August 1840.

As with the 1830-32 Edinburgh-Glasgow scheme, the firm of Grainger & Miller was appointed to engineer the 1835 line, but on this occasion the younger partner, John Miller, was given full responsibility for the project. Miller was by now in his early thirties, having been born at Ayr in 1805. To satisfy the English shareholders, he was obliged to seek from time to time the 'superior' advice of some prominent English engineers, including Joseph Locke, John Rastrick and Charles Vignolles. On most occasions, however, Miller was permitted to follow his own judgement. The route was almost certainly chosen by him, although 'rubber-stamped' by the Englishmen.

During 1835-38 a great deal of time was spent in negotiations with landowners who clearly sought the best possible price for purchase of their property, as well as compensation for loss of amenity, access etc. The vigorous canal opposition had also to be stemmed, but the clear evidence from England was that a system of inter-city railways could now be regarded as an inevitable consequence of the march of progress, so in the final analysis there was little point in trying to swim against the tide. The principles established between Liverpool and Manchester were rapidly extending to other cities and their hinterlands, and so arguments relating to the similarity or dissimilarity between Liverpool/Manchester and Glasgow/Edinburgh were to a large extent redundant. The authorising Act of Parliament was passed on 4th July 1838.

Blood, Sweat and Tears

It was not until the autumn of 1838 that the first contract was let, and the onset of the winter months meant that serious progress was not achieved until the spring of 1839. The work was divided into 21 contracts, two of them devoted entirely to viaduct construction. 7, 000 men were employed, with 700 horses and 12 steam engines.

A large proportion of the labourers employed were Irish, a circumstance which brought special problems, perhaps to a degree greater than in England, where many navvies from the Emerald Isle also found work on railway contracts. Anti-Irish prejudice was already well established among the people of Lowland Scotland, where for several decades there had been a steady influx of labourers from furth of the Irish

Sea into industries such as coalmining, ironstone mining, quarrying, calico printing and brickmaking. Regardless of the position concerning labour supply and demand a feeling persisted that the Irishmen were stealing Scottish jobs, and this led to prejudice, bigotry and open hostility. During Glasgow Fair Week, for example, an activity know as 'Hunting the Barneys' – the taunting and bullying of Irishmen – had become firmly established as an annual tradition.

The Irish labourers employed on the Edinburgh & Glasgow contracts found difficulty in coming to terms with life in the local communities where they were lodged. At Springburn a force of twelve special constables was appointed to maintain peace between navvies and residents. Further out from the city, problems were experienced at Kilsyth, where labourers working on the railway at neighbouring Croy were housed. The situation was described, retrospectively, in the Rev. Robert Anderson's excellent *History of Kilsyth*, published in 1901:

> When the Edinburgh and Glasgow Railway was made and the Roman Catholic Irish invaded the town in hundreds, the native antipathy smouldered deep. For a while it broke out only in taunts and sporadic battles at street corners, but at last it burst out into a big cruel riot, which lasted for hours. The Irish were pursued through the streets, and chased up closes into their houses. They jumped from windows, and everywhere fled, wounded and bleeding.

As well as friction between Irishmen and native Lowland Scots, trouble sometimes developed between Irish and fellow-navvies of Scottish Highland origin. In an incident at Winchburgh, infantry and dragoons were required to take firm action for prevention of bloodshed between large forces of hostile Irish and Highland navvies, each over 1 500 strong.

However, the incident which attracted most attention during this era involved neither Lowland Scots nor Highlanders. This was the murder, on 10th December 1840, of an English ganger named John Green, allegedly by three Irishmen named Dennis Doolan, Patrick Redding and James Hickie, at Crosshill Bridge, east of Bishopbriggs. Doolan had previously worked under Green on a railway contract in Cheshire and grudges may have been borne. Green had been moved by the Edinburgh & Glasgow contractors to the Bishopbriggs sector on 8th December and had immediately taken steps to dismiss Doolan from his employment there. After the murder Doolan and Redding fled, but Hickie was taken into custody and said enough to implicate all three. Doolan and Redding were thereafter picked up in different parts of England and brought back for trial. Little attempt was made to defend them and all three were sentenced to be hanged, although Hickie later had his sentence commuted to overseas transportation. The theatrical procession, bearing the accused from Glasgow to Crosshill on 14th May 1841, and the execution at the scene of the murder, was witnessed by very large numbers of excited spectators, and guarded by several companies of infantry and four troops of cavalry.

Anti-Irish prejudice reared its ugly head throughout the episode and much arrant nonsense was spoken. The sensationalist outpourings in the Scottish press would have done justice to the worst of today's tabloids, although at least they were published in smaller print. Doolan and Redding as a 'natural product of the system which has long existed in Ireland under the popish and O'Connel regime' had 'set about the murder of a fellow creature as if it were a mere affair of course'. This, according to the *Edinburgh Post* was 'not the feeling of Scotsmen or of Englishmen of any class, however oppressed or injured'. The Lord Justice Clerk at the trial was no better, commenting that murder was a crime 'of rare occurrence in this part of the kingdom' although 'by no means rare in another part of her Majesty's dominions, of which those prisoners are natives'. He deplored 'the brutal conduct which is exhibited in so many counties of that land, and the frequent escapes from justice which there take place . . .'

Bitterness towards Scots, harboured by Irish people as a result of this affair, took a long time to subside. It is echoed in a comment made by James Handley in his book on *The Navvy in Scotland*, published in Cork in 1970. Referring to the execution procession he observes:

> One is left with the feeling that only for the Irish would the Scottish authorities have gone to such ceremonial lengths.

The murdered man, John Green, was buried in the Old Aisle Cemetery, Kirkintilloch. His tombstone can be seen in the old, walled part of the graveyard, quite close to the main entrance.

A Flat Level Plane

As previously mentioned the Edinburgh & Glasgow line was engineered on 'Stephensonian' principles, to provide a nearly-level route for embryo steam traction and allow the novice locomotives to show what they could do in terms of speed. John Miller's aim was to avoid gradients steeper than 1 in 880, and the success of this objective is still in evidence today. Between Bishopbriggs and Haymarket there is no greater incline than 1 in 882, apart from a very short one of 1 in 600 at Polmont. To this end he created an impressive infrastructure of viaducts, bridges, tunnels, cuttings and embankments. Although the route via Falkirk is not particularly direct, as the crow flies, between Glasgow and Edinburgh, directness was nevertheless an objective at specific localities. From Falkirk to Winchburgh the railway follows the route of the earlier Union Canal for much of the way, albeit at a slightly lower altitude. Between Polmont and Linlithgow, Miller could have adopted the Canal course, up one side of the Avon Valley and back down the other, but chose instead a more direct route across the valley, at the expense of greater engineering works – a large viaduct of 23 arches (as opposed to 12 on the Union canal aqueduct) and half a mile of embankments.

The Avon Viaduct is nowadays classified as a listed building, Grade A, as indeed is the even more impressive Almond Viaduct, ten miles further east. The 36-arch Almond Viaduct is considered to be one of the finest stone railway viaducts in Britain – comparable with the Ouse Viaduct on the London & Brighton line or the Royal Border Bridge at Berwick. The 8-arch Castlecary Viaduct is also a fine, listed structure, and these three viaducts, taken together, can perhaps be regarded as the most obvious symbol of the pedigree of the Edinburgh & Glasgow Railway as a finely engineered line of the great Railway Age, and an enterprise of the greatest significance.

The concept of a 'Flat, level plane' running the whole way between the two cities is compromised, to an unfortunate degree, by the precipitous descent into Glasgow, at the western end of the line. The topography of the Glasgow approach predetermined that an exception to the 1 in 880 gradient limit would be necessary here, and indeed the advance into Glasgow still begins with a mile at 1 in 809 from Bishopbriggs, but the situation was immeasurably worsened by the attitude of the Forth & Clyde Navigation Company, who steadfastly refused to allow a bridge of any kind to be constructed over their ‘Cut of Junction’ Canal, which joined the Forth & Clyde to the neighbouring Monkland Canal, in the north part of the city. In consequence, it was necessary to burrow beneath the ‘Cut of Junction’, involving not only the construction of an expensive tunnel, but the laying of the railway down a steep gradient (average inclination 1 in 45), for well over a mile, to reach a terminus at Queen Street, Glasgow.

The requirement to make special arrangements for working this Cowlairs Incline put paid to any notion of really fast speeds between the cities. It was considered that locomotives, even working in multiple, would be quite unable to lift trains out of Queen Street Station. A large beam engine was ordered from the engineering firm of Kerr, Neilson & Co., Finnieston, Glasgow, and installed in a stone-built engine house at the head of the incline, with the intention that trains would be wound up the hill from Queen Street to Cowlairs, employing an endless hemp rope for the purpose.

An Outrageous Anachronism

The Edinburgh & Glasgow Railway was laid with malleable iron rails, weighting 75lb to the yard. They were laid down on whinstone blocks, of size approximately 2ft. x 2ft. x 13in., although in certain places transverse sleepers or ‘half logs’ or larch, 9ft. long x 10in. broad x 4 in. thick were employed. The decision to adopt stone blocks, by then decidedly ‘old-fashioned’ technology, was taken at the Third General Meeting of the Edinburgh & Glasgow Railway, held on 27th August 1839, by the Directors of the Company ‘in accordance with the opinions of their eminent Consulting Engineers, Mr. Rastrick & Mr. Locke, as well as of Mr. Miller...’ They were influenced in their decision by the fact ‘that there existed along the line great abundance of basalt or whinstone, of the very best description for blocks...’ The Proceedings of the meeting were published in the *Scotsman* of 7th September, eliciting a knowledgable ‘letter to the Editor’ from a well-known reader:

> Sir, – I observe that at a meeting of the shareholders of the Edinburgh and Glasgow Railway, it has been resolved to use stone blocks as sleepers for the rails. Perhaps the company have decided correctly in coming to this determination, but from the following fact I fear that the contrary is the case. The railway from Dublin to Kingstown was first laid with stone blocks (granite), but it has been found necessary to lift the whole of these and supersede them with wooden beams, which are now placed longitudinally beneath the rails. When I passed along the line lately, I saw the displaced blocks of stone scattered about, and of no value whatever. I was informed that the stone blocks were found unavailing from their unyielding property, which caused immense damage to the vehicles, and that the noise made by the train was altogether unbearable; in short, that it was found absolutely necessary to dismiss the stone blocks and adopt wood in their place, although at a greater expense. At present the train goes in a smooth delightful manner, with very little noise. It is quite possible that the Edinburgh and Glasgow Railway Company may possess some facts leading to the conviction that stone is best for the purpose, but such a fact as that which I relate is surely worthy of their investigation before ultimately committing themselves in this very important matter, – I am, Sir, yours,
>
> William Chambers.
>
> 19, Waterloo Place, Edinburgh.
> 11th Sept. 1839.

The points made by Chambers were valid in the extreme. The troubles experienced on the Dublin & Kingstown Railway (opened 1834), and attributed to the unyielding properties of stone sleeper blocks, had been widespread, varied and costly. The springs and axles of rolling stock, for example, had required frequent replacement. Blocks frequently sank, or tilted, resulting in the breakage of spikes, keys, chairs and/ or rails. By early 1837 almost all the chairs had loosened on their sleepers, thousands of keys had been displaced, and springs of engines and carriages were continually breaking. The decision to substitute sleepers of wood was by then regarded as inevitable, and by December 1839 no stone sleepers remained on the Dublin & Kingstown, the expense of relaying having been £2,440 per mile.

With regard to this matter the Directors of the Edinburgh & Glasgow Company were, however, intransigent. The line was laid with stone blocks for most of its length, necessitating an expensive relaying on wood within a short space of time, just as on the Dublin & Kingstown a decade earlier. On 31st July 1849, for example, the traffic Committee ordered that the stone blocks in Bishopbriggs Cutting be taken up and replaced by sleepers. The displaced blocks were not quite 'of no value whatever' as suggested by Chambers. Being of good square shape they were found to be of particular value for the construction of loading banks for goods traffic, and many examples of this use could be seen along the line. With the decline in local goods traffic in recent years, however, this feature has virtually disappeared. Nevertheless, the whinstone blocks can still be seen fulfilling other useful purposes at many locations. At Croy they form an interesting, decorative kerb at the north entrance to the station; at Linlithgow Union Canal Basin they have been used to face a quayside; at Waterside, near Lenzie, they are in use as stepping-stones across the River Luggie; and frequently they can be noticed as part of the fabric of stout stone walls.

The gauge (i.e. distance between the rails) adopted by the Edinburgh & Glasgow Company was British standard 4ft. 8½ in. This was not necessarily an inevitable decision. It seems fair to state that 4ft. 8½ in. was regarded as something of an eccentric dimension in Scotland. On the Monkland & Kirkintilloch, and several neighbouring lines, it was rounded down to 4ft. 6in. (four and a half feet), while on the Arbroath & Forfar and the Dundee & Arbroath the broader gauge of 5ft. 6in. (five and a half feet) was employed. John Miller took an interest in the 5ft. 6in. option, and actively considered it for the Glasgow, Paisley, Kilmarnock & Ayr Railway, while planning this line during 1837-38. However, it was decided that the GPK & A should share their Glasgow-Paisley section with the Glasgow, Paisley & Greenock Railway, for which the influential English engineer, Joseph Locke, had advised 4ft. 8½ in. This emphasised to Miller the value of a gauge which could be shared with adjacent, connecting lines, enabling through traffic to operate, and brought him into the 'Standard Gauge' camp. If he had any lingering doubts when planning the Edinburgh & Glasgow he would certainly have been kept 'on the straight and narrow' by the various English consulting engineers who had been appointed to guide him.

Along the way the E & G was provided with attractive cast-iron mileposts, reflecting contemporary practice on turnpike roads. At the time of writing the first seventeen of these posts, out from Glasgow (as far as Greenhill) still survive, except Milepost 8 (Waterside) and Milepost 12 (Croy Cutting). They give distances from/to Glasgow Queen Street and *Haymarket* (the terminus from 1842 until 1846). There are no survivals east of Greenhill.

Some of the original stone sleeper blocks, built into a wall beside the railway at Cadder. *D. Martin*

Original milepost at Woodilee, east of Lenzie. *D. Martin*

Promotion Par Excellence

Nowadays, business promotions are planned precisely, with careful calculation of the likely return on capital expended. During the early Railway Age grand spectacles were organised according to the custom of the times without any apparent consideration of their commercial value. The ceremonial opening of the Garnkirk & Glasgow – a coal railway – was stylishly staged in September 1831 without any clear intention of generating passenger revenue. The railway company was then taken by surprise when Glaswegians in large numbers displayed great interest in train travel – for its own sake – along this, the city's first public railway. In 1832, to provide an amenable destination for these enthusiastic excursionists, it was obliged to provide the world's first railway-owned catering establishment – Gartsherrie Inn – in the midst of the coalfield at the 'country' end of the line.

A decade later the income-generating value of such events was without doubt better understood. Curious onlookers could clearly be cossetted as potential customers. Furthermore, a tradition had already been established in Scotland of targeting the entire social spectrum, without the same emphasis on the upper middle classes displayed by such English railways as the Great Western and the Grand Junction. The well-documented banquet to mark the official opening of the Edinburgh & Glasgow Railway on 18th February 1842 was provided for 'those and such as those', but ordinary folk had not been forgotten by the E & G Directors. A few weeks earlier on 1st and 3rd January, when 'Ladies and Gentlemen' were admitted to the whitewashed and gas-lit 'Great Railway Tunnel' at Queen Street for 1s per head, 'Trades People' were allowed in for 6d. On Wednesday 5th January another opportunity to visit the tunnel was provided at a further reduced rate. Falkirk Tunnel, likewise illuminated, was thrown open for inspection at the same time as Queen Street Tunnel. On 17th February, the eve of the banquet, the public at large were invited to Queen Street Station to inspect the preparations. They were able to gawp at the huge curtains which screened off the 'country' end of the train shed, the boarded-over platforms, the tables set for 1,200 people, the augmented gas illumination, and the gas jets forming the outline of a locomotive. On the great day itself, it they so wished, they could join the crowds at the lineside to cheer the trainful of Glasgow guests and the trainful of Edinburgh guests. Some of the fun, of course, was reserved for the privileged – seats on the trains (all First Class), the food, the speeches, and the broken incline rope suffered by the last train of the day.

Class Distinctions

John Willox's *Guide to the Edinburgh and Glasgow Railway*, published during 1842, the railway's inaugural year, lists four passenger trains in each direction. 'All stations' services left Edinburgh at 7.00am, 3.00pm and 5.00pm, and Glasgow at the same hours. 'Express' services, stopping only at Linlithgow, Falkirk and Castlecary left from each terminal at 11.00am. In addition, a 'Train of Third Class Carriages' left each end 'along with the Luggage Train at 6 o'clock each morning'. Time taken between the cities was 2¼ hours by the 11.00am express services and 2½ hours by the stopping trains. Schedules for the 6.00am 'luggage' trains were not given. Intermediate stations were provided, in the first instance, at Corstorphine, Ratho, Winchburgh, Linlithgow, Polmont, Falkirk, Castlecary, Croy, Kirkintilloch (i.e. Lenzie) and

Bishopbriggs. Fares between Edinburgh and Glasgow were 8s First Class, 6s Second Class and 4s Third Class. A concession fare of 2s 6d was available to Third Class passengers willing to travel on the 6.00am 'luggage' train. There were Sunday Trains from each end at 7.30am and 5.30pm, stopping at all stations.

Two years later, according to Henry Tuck's *Every Travellers Guide*, the fares were still the same. Additional stations were by this time provided at Gogar, Broxburn and Camelon, but no mention was made of 'Corstorphine'. There were now six passenger services in each direction, with starting times between 7.30am and 10.00pm. Only one of the down trains (the 7.30am ex-Edinburgh) and two of the up trains (the 7.30am and 6.30pm from Glasgow) stopped at all stations. The 10.00pm was an express, in the 1842 sense, stopping only at Linlithgow, Falkirk and Castlecary. The stops made by the others varied greatly from train to train. The fastest trains (actually the 1.00pm from either end) now took exactly two hours. There were still two Sunday services each way, described as follows:

> These trains start from each end at 8 o'clock morning, and 5 o'clock afternoon calling at all stations, and departing therefrom at the same time as the daily 6 o'clock evening Train, making allowance for the difference of the hours of starting at each end.

With regard to the week-day trains the following supplementary note was given:

> All the Trains carry First, Second and Third Class Passengers, except the 1 o'clock and 10 o'clock p.m. Trains which carry first and second class only – the 1 o'clock, however, carrying third class betwixt Glasgow and Kirkintilloch. By the 7½ o'clock morning Train, the Fare, in Stand up Carriages, for the whole distance, is only 2/6. In that Train there are also Third Class seated carriages, the Fare by which is 4s.

The anomaly whereby there existed two grades of 'Third Class' passenger accommodation, seated and standing, at different fares, was resolved the same year. The 'Winter Arrangements 1844-5', introduced on 1st November 1844, incorporated a designated Fourth Class, with a fare of 2s 6d between the cities. The fares for First, Second and Third Class passengers remained at 8s, 6s and 4s, respectively. With supreme irony, however, a new variation in Third Class travelling conditions was introduced at exactly the same time. From 1st November 1844, Third Class passengers by the new 1.30pm all-stations trains, in each direction, travelled in carriages with 'Seats and Covers' at a special fare of 3s 10d between the cities. This was to comply with the famous 'Gladstone's Act' for regulation of railways, which came into force on 1st November, stipulating that all passenger railways should each day run at least one train along their lines from end to end, in each direction, at a speed of not less than 12 miles per hour, including stoppages at every station, and employing carriages provided with seats and protected from the weather, to the satisfaction of the Board of Trade – fares in these trains not to exceed one penny per mile. The trains so designated soon came to be known as 'Parliamentary' trains, in consequence of which there were, in effect, five classes of passenger on the Edinburgh & Glasgow Railway – First, Second, Third, Fourth and Parliamentary.

The cheapest accommodation on the Edinburgh & Glasgow Line – whether roofless and seatless Third Class 'stand-ups', or the later Fourth Class – was introduced partly

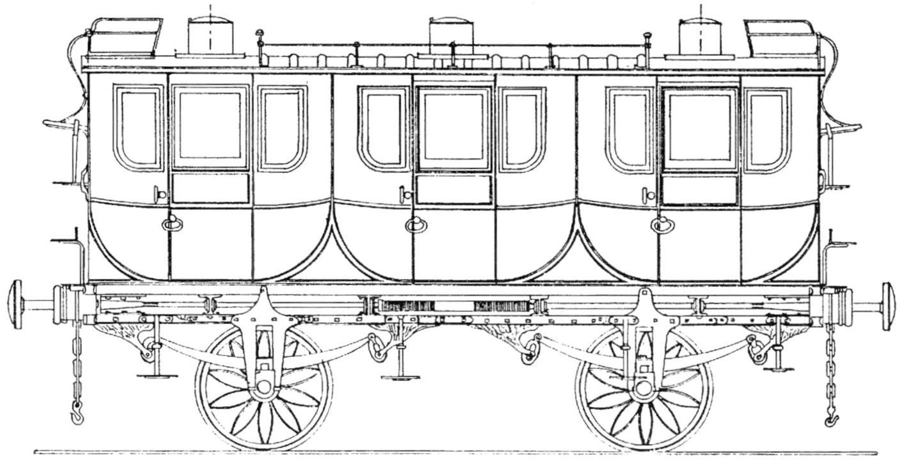

First Class Carriage – with roof, glazing, and upholstered seats. ***A.A. Maclean Collection***

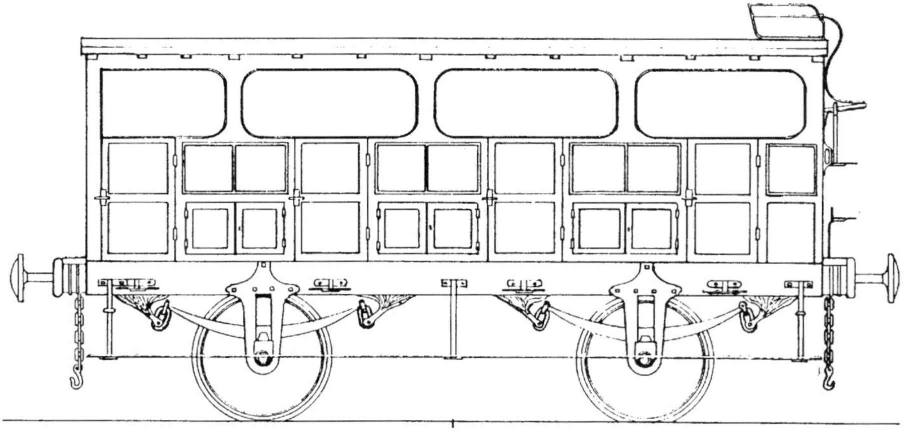

Second Class Carriage – roofed and seated, but with no glazing. ***A.A. Maclean Collection***

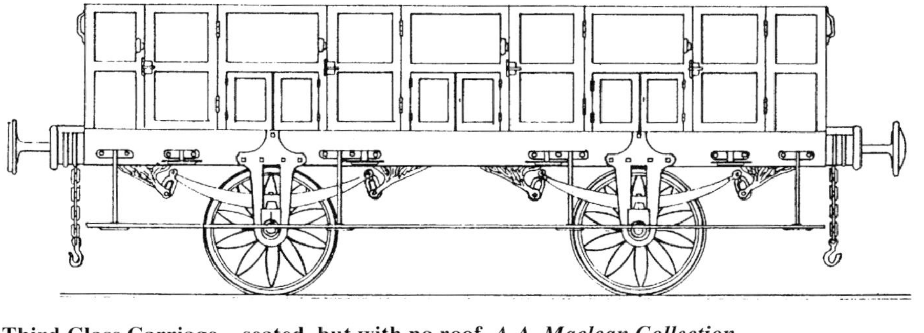

Third Class Carriage – seated, but with no roof. ***A.A. Maclean Collection***

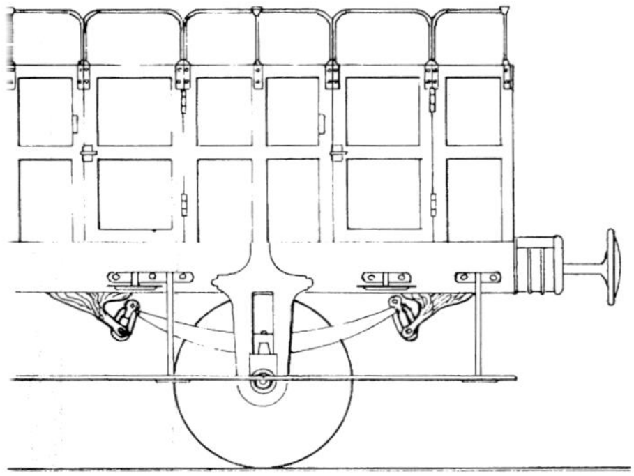

Drawing of one end of a seatless 'Stand-up' carriage – originally designated 'Third Class'. ***A.A. Maclean Collection***

to win passengers from the canal route between the cities, which had offered a dramatic reduction in fares when the Edinburgh & Glasgow Railway was opened to public traffic. It soon became clear that the much shorter journey times possible by the railway route were in themselves sufficient to gain the patronage of all but the most impoverished passengers. Nevertheless, the additional passengers won over by the introduction of 'rock-bottom' fares were certainly appreciated by the Edinburgh & Glasgow Company, and consideration was given to a 'cheap fares' policy for its own sake.

One problem was how to persuade members of the different social classes to occupy appropriate sections of the trains. The First Class accommodation, although bearing a tariff only $33\frac{1}{3}\%$ above Second Class, was appreciably more comfortable than any other, with spacious, transverse non-corridor compartments, provided with well-upholstered seats (three facing forwards and three backwards) and glass windows. The Second Class vehicles had seats and roofs but no glazing (and in consequence were very draughty), most of the Third Class vehicles had seats, but were completely open above waist level (and therefore roofless), while the original Fourth Class stock had neither seats nor roofs. The railway company clearly hoped that the 'comfort-gap' between First and Second Class accommodation would induce all persons of reasonable substance to travel First Class, but in practice this was far from the case. Open stock was popular with all social groups, especially in warm, dry weather. Manufacturers, merchants and businessmen could all be encountered in cheaper sections of the train, while it was even claimed that Glasgow *magistrates* were quite happy to travel in 'stand-up' carriages. Apart from the provision of luxury accommodation, the railway company adopted other means of persuading the well-heeled to travel First Class. According to 'A DAILY SUFFERER' in the correspondence columns of the *North British Railway & Shipping Journal* of 3rd October 1846 'the delivery of the second, third and fourth-class tickets' at Glasgow was entrusted to 'one solitary boy' while often 'the man at the first-class table' stood looking on with his hands in his pockets. The editor suggested that the Second Class passengers should be allowed to 'enter along with the first' – but doubtless the railway company would have considered this a dilution of First Class privilege.

During inclement weather, of course, travel in open Fourth Class 'stand-ups' was quite intolerable. A paragraph on the subject in the *North British Railway & Shipping Journal*, of 11th September 1847, is quite scathing in its language:

> ABOMINATION OF FOURTH-CLASS CARRIAGES
>
> That inclement period of the year is fast approaching which tells so fearfully in our mortality returns. In another month we will have winter gales and torrents, with their concomitants in the shape of coughs, colds, consumptions, and catarrhs playing havoc amongst our gentry, and decimating our labouring population. To see our ill-fed, ill-shod, and ill-clad artizans shivering and shrugging, in consequence of the cold and wet, as they pass along the streets, is a scene, in all conscience, heart-rending enough; but if any person wishes to see misery in that shape to its fullest extent, he has only to take a trip in a fourth-class carriage of the Edinburgh and Glasgow Railway, and there he will behold it to his heart's content. If the benevolent individual, who may wish to obtain such knowledge, however, shrinks from the exposure himself, he will perhaps obtain accommodation along with the railway officials in a back coupee, and thus he will have an opportunity of observing, without suffering damage. The Edinburgh and Glasgow Company enjoy the privileged distinction of being the only railway in Scotland that runs fourth-class open stand-up carriages; and they have, with a perseverance worthy of a better cause, doggedly stuck to that system of refined cruelty in spite of all that can be said against it. The directors have a great regard for the population in a moral and religious point of view; but in a personal and physical one, they do not care a pin-point.

However, travel in open, roofless carriages had its bizarre, and even comical side. An old man, Mr. John Paul, sent some interesting reminiscence to the *London & North Eastern Railway Magazine* in 1927:

> I was greatly interested in the recent articles about the beginning of the Glasgow to Edinburgh Railway in 1842. I was born in Polmont in January of that year, and can remember as an infant being lifted in my mother's arms over the wall on the bridge at the station and saw my father, who was a passenger in a third-class carriage from Edinburgh, who waved his umbrella to us. The third-class carriages then had neither seats nor covers, and elderly ladies brought stools with them.

A report in the *Glasgow Argus* of 24th February 1845 is reminiscent of some of the stories told about the Liverpool & Manchester Railway during the early 1830s, before passengers had learnt about the dangers of trying to join or leave a speeding train:

> On Wednesday morning, as the seven o'clock train from Glasgow was approaching the old Kirkintilloch station, the hat or bonnet of a passenger by one of the open carriages, was blown off his head. Being tipsy at the time, he attempted to open the door of the carriage to get out after his head-piece, but was prevented by the other passengers. Seemingly determined, however, not to be controlled, he threw himself over the side of the carriage, but after doing so he continued to hold on by the buffer-end till the train was passing the now disused station mentioned above, when he relinquished his grasp, and tried to get a footing upon the platform at this place; but almost, as a necessary consequence, on account of the speed at which the train was proceeding, he was thrown down with great force, and was much injured about

> the face and head. He was brought to Glasgow during the forenoon, however, and sent to the Royal Infirmary, where his wounds were dressed, and we are glad to learn, that from present appearances, hopes are entertained of his recovery. Nobody seems to have been in the least to blame but the foolish man himself.

The 'old Kirkintilloch station' had been in use between 1842 and 1844, when a new station was provided on a site further east. The old site was re-occupied in 1848, however, and is now known as Lenzie Station.

Managing the Mundane

The Edinburgh & Glasgow was promoted primarily as a passenger line, with perhaps a tendency to underrate its potential as a freight route. Its asset of speed was not regarded as fundamental in this market, and it was considered that the Forth & Clyde/ Union inter-city canal route would continue to enjoy a major share of the goods traffic. Furthermore, during the early 1840s the Forth & Clyde Canal, in particular, was 'better connected' than the Edinburgh & Glasgow Railway. The railway existed in isolation, without any connecting links to other lines, while coasting vessels could leave the Canal and proceed to ports on any part of the British coastline, to Ireland, and even to continental Europe. The Forth & Clyde Canal even had superior links with the important system of mineral railways in the Monklands, with excellent loading facilities at Kirkintilloch, even extending to 'waggon-boats' whereby railway wagons could be taken along the canal in vessels fitted with rails and turn-plates. The Edinburgh & Glasgow constructed a junction-line to the M & K at Garngaber, near Lenzie, in 1844, but until uniformity of gauge could be achieved it was useless as a through route for goods and minerals.

It soon became clear that the Monkland lines would be forced to change their gauge from 4ft. 6in. to standard 4ft. 8½ in., and with this in mind a scheme was devised for the Edinburgh & Glasgow Railway management to assume control of numerous railway and canal undertakings in central Scotland, including the Monkland & Kirkintilloch, Ballochney, Slamannan, Wishaw & Coltness and Scottish Central railways, and the Forth & Clyde, Union and Monkland canals. The scheme was clearly geared to the requirements of goods and mineral traffic, much more so than passengers, and could be regarded as a major change of emphasis by the E & G management. In anticipation of the success of the scheme, the E & G took charge of all rolling stock belonging to the M & K, Ballochney, Slamannan and Wishaw & Coltness railways and assumed responsibility for working these lines on 1st January 1846. A high-profile manager, Bryan Padgett Gregson, was appointed on 1st February. The scheme fell through, however, because of suspicion of Scottish motives on the part of English shareholders of the E & G, and management of the mineral lines was returned to their proprietors.

Unlike so many other early Scottish lines, the Edinburgh & Glasgow was not promoted as a 'coal railway', recognising that the adjacent canals were in a strong position with regard to the transportation of minerals in bulk. Furthermore, the point was made that by choosing the level route by way of Falkirk the principal mineral field was skirted rather than intersected, thus reducing the potential for traffic in coal and other minerals. In fact there were substantial mineral deposits close to the chosen route. In Polmont parish alone there were already important collieries in production at

Redding and Middlerig, together employing over 400 men. Both splint and soft coal were mined, and sent mostly to the Edinburgh market. The larger colliery, at Redding, was owned by the Duke of Hamilton. Coal from the pits was raised by steam winding-engine and conveyed by private railway down inclined planes, to the Union Canal for loading. The Edinburgh & Glasgow Railway Company may at first have been willing to yield precedence in this traffic, and indeed Willox's *Guide*, published for the opening of the E & G in 1842, made mention of the Duke of Hamilton's coal from Redding Colliery being 'shipped for the Edinburgh market' along the Union Canal, but the concession was of short duration. On 10th October 1842 an Edinburgh coal merchant, James MacNaughton, notified the Edinburgh public that he had 'taken advantage of the facilities afforded by this Railway for bringing into Town, in a fresh state, the COAL of the Redding Coalfield'. He was confident that 'his Establishment, at the Railway Depot, Hay Market' had only to be tried 'to be found to supply an article superior to any yet brought into Edinburgh'. In the event, Redding Colliery became a major source of mineral traffic for the railway, over a lengthy period of time.

A Forgotten Genius

Few would disagree that, for a small nation, Scotland has produced a remarkable number of outstanding inventors, responsible collectively for a very respectable proportion of the necessities of modern life. Some of these inventors are accorded their deserved recognition, but others are almost forgotten – like Alexander Bain, an electrical innovator who was born at Watten, Caithness, in 1810. Inventions credited to him include the electro-chemical telegraph, an electromagnetic printing telegraph, the earth circuit, a single-needle telegraph, an electromagnetic engine, an electric fire alarm, electric sounding apparatus for ships, and various types of electric clock. He has even been credited with the groundwork for the invention of the fax machine! (See the *Scotsman* of 14th September 1991). During 1845-46 he erected a telegraph line along the whole distance of the Edinburgh & Glasgow Railway, at a cost of only £50 per mile (one-third of the charge per mile then being made by his famous competitors, Cooke and Wheatstone). He then put it to novel use, as reported in the *Practical Mechanic & Engineer's Magazine* of May 1846:

> Since the successful introduction of Bain's single wire electric telegraph, the ingenious inventor has directed his attention to the contrivance of a plan whereby a system of uniform time may be established throughout the country. This idea he has most successfully worked out by means of an electric communication between the places in which it is proposed to maintain that system. In order to ascertain, by actual trial, the value of this scheme, Mr. Bain placed the pendulum of one of his electric clocks at the Edinburgh station of the Edinburgh and Glasgow railway; and the works of a common time-piece at the Glasgow terminus were set in motion simultaneously with it, by means of the electrical wire connecting the two stations. The beat of the pendulum being also exhibited on a dial beside it, showing the movement as it was simultaneously made at the Glasgow station.
>
> We look upon this invention as being one of the greatest importance, as by its introduction the great evil of variation of time in distant situations will be entirely avoided, for by establishing a chain of electrical communication from the Land's End to John O'Groats, the exact solar time may be given with the greatest possible degree of precision in both.

This of course was an era when there was no uniformity of time, and indeed great variation between different parts of the country. Permanent synchronization of time between Glasgow and Edinburgh was achieved in January 1848, when both cities agreed to adopt 'Greenwich Time'. By this date, regrettably, Alexander Bain's telegraph wire between them had fallen into disuse. On 4th March of that year the *North British Railway & Shipping Journal* reported that it had become quite rusty – its batteries were uncharged, its wires broken, and it was altogether 'out of joint'. To replace it a new one was being erected:

> It consists of five wires, which are insulated in a very complete and beautiful manner, so as not to divert the fugacious element on which so much depends, in the slightest degree. One wire, we believe, will belong entirely to the Telegraph Company, who will transmit their messages along it independently of the railway and everyone else. Another wire will be used for direct communication on railway business betwixt Edinburgh and Glasgow, and the rest for communication betwixt the stations and the termini. When completed, Glasgow will be in direct telegraphic contact with London, Dover, Portsmouth, etc.

Alexander Bain died on 2nd January 1877, at Broomhill Home, Kirkintilloch. He is buried in the Old Aisle Cemetery, Kirkintilloch, not far from the murdered ganger John Green (of the 'Doolan & Redding' affair), but in a more modern part of the cemetery.

A Thundering Divine

In his famous book *The Springburn Story*, the late John Thomas referred to the circumstance when a train 'was met at Edinburgh by a thundering divine who ranted at the detraining passengers, and reminded them that they had bought tickets to Hell'. The occasion was the running of the first Sunday train on the Edinburgh & Glasgow Railway, on 13th March 1842, and the 'thundering divine' was the Rev. William Chalmers Burns of Kilsyth. A contemporary journalist in the *Scotsman* employed another descriptive phrase, somewhat less polite – 'that unhappy fanatic' – but W.C. Burns would not have cared. He was a totally committed evangelist, of towering presence, who purveyed his Sabbatarian message in many other places, at many other times. In August 1841, for example, he was to be found in Newcastle, lambasting the Newcastle & Carlisle Railway for advertising a Sunday excursion from Newcastle to Carlisle: 'A REWARD FOR SABBATH BREAKING. PEOPLE TAKEN SAFELY AND SWIFTLY TO HELL, NEXT LORD'S DAY, BY THE CARLISLE RAILWAY, FOR 7s.6d. IT IS A PLEASURE TRIP! *The Devil will give a cheaper Passage to poor People by the River Boats*'. Later on Burns became a missionary in China, which accounts for the fact that his portrait depicts him in oriental garb.

The Rev. Burns was not the only enthusiastic Sabbatarian who harangued the Edinburgh & Glasgow Railway for running Sunday trains. There was Robert Haldane, who published various editions of his pamphlet *On the purposed desecration of the Sabbath by the Directors of the Edinburgh and Glasgow Railway*, Robert Murray McCheyne who prepared *A testimony against the running of railway trains on Sabbath* (1841), R.T. Sawers who compiled *Queen Street: a poem, in seven parts, with illustrative notes of all the principal speeches and proceedings in the great contest for the keeping of the Sabbath against the prevailing sin of Sabbath trains, down to the present time*

(1848), and Sir Andrew Agnew of Lochnaw, who purchased shares in the Edinburgh & Glasgow Railway in order to be able to speak out on the subject at company meetings. John Leadbetter, who had a Sabbatarian conscience, resigned as chairman over the matter. Many meetings were held and a great deal of campaigning went on, but what finally stopped the Sunday trains was an 'unholy alliance' between the Sabbatarians and mercenary English shareholders in search of greater dividends. The existing Board was toppled from office and the Sunday services were all discontinued, with effect from 1st November 1846. They did not resume until the Edinburgh & Glasgow Company was absorbed by the North British Railway in 1865.

English Interference

When in 1845 the scheme was hatched for a broad-based amalgamation between the Edinburgh & Glasgow Railway, the Scottish Central, the various Lanarkshire mineral railways and the Forth & Clyde, Union and Monklands canals, there was consternation amongst the merchant communities of Glasgow and Edinburgh. They immediately sensed the danger of a grasping monopoly which would corner the market and ensure that transport costs in the Central Belt were maintained at a permanently high level.Viewed from a distance of two hundred miles the standpoint of the Lancashire proprietors of the Edinburgh & Glasgow was somewhat different. They harboured a deep-seated suspicion of Scottish-based shareholders of the E & G, believing that as local businessmen they would be quite keen to keep transportation costs in Central Scotland down, and would not capitalize on any monopoly situation by imposing high tariffs. The prime object of their suspicion was John Wilson of Dundyvan, the powerful Coatbridge ironmaster, who wore one hat as a director of the Edinburgh & Glasgow Railway Company, but another as a militant local trader, constantly campaigning with his fellows for lower rates. Wilson's known view was that if the proposed amalgamation went ahead it should be on the basis that the enabling legislation would incorporate a 'maximum tariff' clause.

On 3rd July 1846 the 'Select Committee on Railway Group No. 61' in London declared the preamble of the bill for amalgamation of the Monkland railway lines with the E & G 'not proved'. The members of the Committee considered that the railway system of the district was not yet sufficiently developed to enable the Committee to decide on the mode of amalgamation proper for it. Despite this setback the E & G directors decided to press on with the amalgamation schemes. The English shareholders had by this time seen enough, however, and on 22nd September the entire Board was forced to resign. John Learmonth, who had succeeded John Leadbetter as chairman in 1842, was replaced by Peter Blackburn. The new directors who took over still included a majority of Scottish residents, but English influence over them was greatly strengthened.

A Line for Leisure

The 5-mile Campsie Branch was opened from Kirkintilloch Junction (later Lenzie Junction) northwards to Milton of Campsie and Lennoxtown on 5th July 1848. A major attraction was the sylvan dell of Campsie Glen, famed as a place of recreation for working people since an altruistic local landowner, John McFarlan of Ballencleroch, had thrown his private share of the glen open for that purpose, some years earlier. An important added attraction was William Muir's well-known public house, the Crown Inn, situated at the foot of the glen. It mattered not that excursionists had almost another two miles to cover on foot, after de-training at Lennoxtown. The extensive goods and mineral traffic on the branch embraced calico printing works, a chemical manufactory, coalmining and limestone quarrying. An excellent description of the branch, during its inaugural month, was published in the *North British Railway & Shipping Journal* of 15th July 1848:

> In our last we copied from another paper a paragraph announcing the opening of the Campsie branch of the Edinburgh and Glasgow Railway; and on Saturday last we had the pleasure of passing along this very important adjunct to the line, and which, without pretending to be prophetic, *at present*, we can venture to say will prove a prolific feeder to the main trunk, abounding as the country does through which it passes in trade and manufactures, and also in population. Amongst other concerns on the line, there are the extensive printing and bleaching works of Messrs. Dalglish, Falconer, and Co.; the secret-works of Mr. Macintosh; the printfield of Mr. M'Nab, at Lillyburn, &c, &c. – The scenery along the new branch is most interesting, combining as it does the grand with the beautiful; the splendid hills of Campsie, the finely-situated castle of Lennox, the gurgling burn, the shady grove; but above all, and within a mile or so of the station, the far-famed Campsie Glen, one of the most charming spots in the world. What an opportunity now for viewing this fairy dell on a fine Saturday, or any day, and of partaking of the good cheer at Mr. William Muir's, whose comfortable Hotel (the Crown) is situated at the very entrance to the Glen, and who, without any exception whatever, we can pronounce (after nearly forty years' acquaintance) to be one of the most respectable and conscientious landlords in Scotland.

In the same issue of the paper 'Cheap Excursion Trains to Campsie' were advertised for Glasgow Fair Week, in addition to regular trains, leaving Queen Street Station at 1.30pm on 15th, 17th and 18th July, returning from Campsie (i.e. Lennoxtown) at 8.30pm on each of these evenings. Return fares were 2s (First Class), 1s 6d (Second Class) and 1s (Third Class).

The First Commuters

The station now know as Lenzie has been named 'Kirkintilloch', 'Kirkintilloch Junction', 'Campsie Junction' and 'Lenzie Junction' at different times. The adjacent community has developed during the lifetime of the railway as an archetypal commuter suburb. This development was intended to begin on 'Day One', since an advertisement appeared in the Glasgow press in August 1841, six months before the opening of the railway, for 'DESIRABLE VILLA GROUND NEAR GLASGOW TO BE FEUED ... contiguous to the Kirkintilloch Station on the Edinburgh and Glasgow Railway'. It was claimed that when the railway was in operation the lands in question would be 'reached in about one quarter of an hour from George Square'. There were no takers at this time. However, when the station was moved back to its regular location from the temporary site it had occupied during 1844-48, building began in earnest. From this date onwards the 'commuter' houses can virtually be numbered, 1-2-3-4, and so on, in chronological order. The first three can still be identified, running southwards along the right-hand side of the road which passes under the railway at the eastern end of the station platforms. Around 1850 the Edinburgh & Glasgow directors gave a boost to this development by announcing the company would offer free 'Villa tickets' to persons building substantial houses near the stations, at the rate of one years free travel for each £100 of the house's value, provided that it was worth at least £500. The

TO BUILDERS.

The DIRECTORS of the EDINBURGH and GLASGOW RAILWAY COMPANY are prepared to grant

FREE TICKETS

TO PARTIES WHO MAY

BUILD

VILLAS

within a Mile of any of the Stations on the Line, not being Town Stations, on the following terms :—

Certified Cost of Villa.	Free Ticket for Years.
£500	5
600	6
700	7
800	8
900	9
1000	10

and so on, an additional year for every additional £100 in the Cost of a Villa.

TICKETS TRANSFERABLE TO A YEARLY TENANT, IF THE HOUSE IS NOT OCCUPIED BY OWNER.

rther Information may be known on application at Company's Offices, George Square, Glasgow.

'Villa Tickets' notice, describing the unique offer available to commuters of the 1850s. *Collection of the late J.F. McEwan*

offer remained on the table for a decade or so. Some of the houses built under the scheme can be readily identified, especially in Garngaber Avenue, a short distance to the north of the railway. They include the large 'castellated' building known as 'The Tower' which dominates the brow of the hill, just beyond Lenzie Old Parish Church.

Lenzie was not the only locality to benefit from the 'villa tickets' offer. On the down side of the line at Bishopbriggs, for example, some houses built under the scheme can be identified in Springfield Road. In 1855 the railway company took steps to clarify that 'season or house tickets west shall not be given between Polmont and Edinburgh, or East between Polmont and Glasgow'. This would appear to suggest that during the early years of the scheme it might have been possible, in theory at least, to obtain a villa ticket for free travel between Bishopbriggs and Edinburgh.

The N.B. and The Caley

As mentioned previously, the Edinburgh & Glasgow Railway existed in isolation during its earliest years, without physical connection to any other line. Within the short period of eight years, however, it was connected comprehensively to a vast system of main line railways, extending from the south of England to Aberdeen.

The first important connection was made in 1846, when the Edinburgh & Glasgow Company extended its own line from Haymarket eastwards through Princes Street Gardens, to link up with the newly-arrived North British Railway at the location now known as Waverley Station. The work was put in hand well in advance, as reported in the *Edinburgh Advertiser* during September 1844:

> The works in connection with the various railways are proceeding with great activity. The drift tunnel through the Mound is nearly completed, and, in the eastern garden, the ground is nearly all levelled for the rails of the Edinburgh and Glasgow Extension Line. Three large old dingy tenements close under the Calton-hill, between Burns's Monument and the Jail, have been demolished, on the North British Line. The brewery property to the west of them will speedily share the same fate, as well as many ruinous and closely wedged domiciles in this plebeian district; by the removal of which the general health of the city, if it do not gain, will undoubtedly lose nothing. In addition to the ruin to which the imperious claims of

Locomotive of the Scottish Central Railway, which formed an important connection with the Edinburgh & Glasgow, from 1848 onwards. ***D. Martin Collection***

> the Railway Acts are consigning so many of the old-fashioned architectural embellishments of our city, it is satisfactory to know that a good deal of building work is going on in various directions. We trust we may take this as a token of a return to the prosperity which this capital enjoyed years ago but which it has lost for some time, for want of a proper stimulative power.

The North British Railway's main line, between Berwick and Edinburgh, was officially opened on 18th June 1846, the Edinburgh & Glasgow connection through Princes Street Gardens to Haymarket on 1st August. Through running into England became possible on 10th October 1848, when a temporary timber bridge over the Tweed at Berwick came into use, pending completion of the Royal Border Bridge, which was in turn opened on 29th August 1850.

An important junction with the Scottish Central Railway was established by the Edinburgh & Glasgow Railway at Greenhill Upper on 22nd May 1848, and three months later the Scottish Central made connection with the Caledonian Railway nearby at Greenhill Lower, on 7th August 1848. Physical connection was thereby formed with lines which extended right down to London, by the West Coast Main Line route. The Scottish Central Railway extended northwards to Perth, and various connecting railways soon took the route onwards to Aberdeen, which was achieved on 1st April 1850. Through running from Edinburgh to Stirling, Perth and the north, was enabled by the opening of the Stirlingshire Midland Junction Railway, from Polmont to Larbert, via Falkirk Grahamston, on 1st October 1850, thus completing a Polmont-Larbert-Greenhill triangle.

During 1847 the various Monkland lines all changed their gauge from 4ft. 6in. to standard 4ft. $8^{1}/_{2}$ in. and through running to and from the Edinburgh & Glasgow Railway was immediately possible by means of the junction at Garngaber, just east of Lenzie. A further connection with this system of mineral railways was established when the short Slamannan Junction Railway was opened, from the Edinburgh & Glasgow Railway near Manuel to the Slamannan Railway at Causewayend, during the same year.

So during a period of four short years, 1846-1850, the Edinburgh & Glasgow Railway was transformed from a wholly isolated route into one that was fully connected to systems stretching over most of the length and breadth of Great Britain. The potential for goods and mineral traffic, in particular, was enormous.

Political Fixes

On 30th May 1991 a special train hauled by Class 91 electric locomotive No. 91019 *Scottish Enterprise* ran from Edinburgh to Glasgow Central via Carstairs, and back by the same route. This was to formally preview the new electric passenger service on this route, operated by *INTERCITY* (unlike the route via Falkirk, which is operated by ScotRail) as an extension of its East Coast Main Line electric service between Kings Cross and Edinburgh Waverley. At a ceremony in Edinburgh, immediately preceding the run, speeches were delivered by Dr. John Prideaux, Director of *INTERCITY*, and Malcolm Rifkind, the Secretary of State for Transport. Mr. Rifkind welcomed the possibility of inter-sector competition between the cities, and looked forward to the day when there would be competition 'not only on service and frequency but also on speed and price'. His speech led to a great deal of premature media speculation concerning the possibility of a fares-cutting price war between the sectors for Edinburgh-Glasgow custom. This possibility seems remote, although the circumstance of two different British Rail sectors operating trains between a pair of cities, on separate routes, is certainly unique.

If a price war does break out, between the Falkirk and Carstairs routes, it will be a case of history repeating itself, for this was exactly what happened in 1848, when the Caledonian Railway opened its route from Carlisle to Carstairs, with forks from the latter place to both Glasgow and Edinburgh. In theory the competition was very unequal, for the Caledonian route was eleven miles longer than that of the Edinburgh & Glasgow Railway, and hillier besides, but the Caley entered the market by introducing *non-stop* trains between the cities, in two hours. The Edinburgh & Glasgow introduced competing expresses, one of which took the journey time down to 1½ hours. Fares came down to 4s (First Class), 3s (Second Class) and 2s 6d (Third Class), much to the annoyance of English shareholders. Before long a truce was called and fares were allowed to return to more realistic levels.

One way to eliminate competition between the Caledonian and Edinburgh & Glasgow railways was to amalgamate them, and exactly that was proposed, on several occasions during the 1850s and early 1860s. A 'Tripartite Agreement' on railway traffic was made between the Caledonian, Edinburgh & Glasgow and Scottish Central railways in 1862, followed by the 'Thirty Years Agreement' in 1863, the provisions of which were intended to replace the 'Tripartite Agreement' after 1866. Meanwhile a Bill for amalgamation of the three railways was introduced, but was rejected by Parliament in February 1864, on Second Reading, without a division. Alarm bells were already ringing in the Edinburgh & Glasgow board room. It was clear that Central Government was hostile to the monopoly transport situation which would be created in West Central Scotland if the Edinburgh & Glasgow and the Caledonian were amalgamated, and that this hostility was unlikely to recede. On 19th May 1864 the E & G board came to the conclusion that it would be pointless to carry on with the Amalgamation Bill, 'the feeling in the House of Commons being so strongly opposed to it', and instead the company should 'at once endeavour to come to terms with the East Coast Lines, and form an alliance with them leaving the Caledonian to amalgamate with the Scottish Central Railway'. This policy was carried to its logical conclusion the following year, and the Edinburgh & Glasgow Railway Company was amalgamated with the North British Railway Company on 1st August 1865, under an Act passed on 5th July, the Scottish Central passing into Caledonian Railway ownership at the same time.

Minerals & Manufactures

Mention has already been made of the important Redding Colliery, owned in the early years by the Duke of Hamilton, but later worked for a long period by James Nimmo & Co. Other collieries were established, very visibly, at the lineside, including Wester Gartshore, Gartshore and Dullatur, also Meadowbank and Carmuirs on the Stirlingshire Midland Junction line. A short distance along branch or connecting lines were numerous collieries, impossible to list comprehensively – Cadder, Wester Auchengeich, Twechar, Bantaskine, Callendar, Blackbraes, Manuelrigg, Craigend, Kinneil – to name but a few. In retrospect, the mineral traffic of the Edinburgh & Glasgow suffered but little on account of the level route chosen – the known mineral fields may have been avoided, but others were found to replace them.

The well-known fireclay seams of the Glenboig/Garnkirk locality in Lanarkshire were found to dip downwards to the east of these places, but they came towards the surface again in the Bonnybridge/Castlecary area, where numerous fireclay mines were established, with adjacent firebrick works. It is said that the Bonnymuir Fireclay & Grateback Company was founded at Low Bonnybridge as early as 1839, but most of the significant firms were established during the 1870s, 1880s and 1890s, or a little later, in the Castlecary-High Bonnybridge-Rough Castle locality, with sidings leading from the Edinburgh & Glasgow railway line. They included John G. Stein & Co., James Dougall & Sons, The Bonnybridge Silica & Fireclay Co., Dykehead Ganister & Fireclay Co., Campbell & Co. and James Anderson & Co., with works at Milnquarter/Castlecary, Bonnyside, High Bonnybridge, Dykehead, Rough Castle and Glenyards, respectively. Of these, only Dougall's works at Bonnyside survives, now in the ownership of Dyson Refractories Ltd. Elsewhere, there was J. & A. Towers' Grahamston Firebrick works, on a branch leading from the Stirlingshire Midland Junction line at Falkirk, and another large firebrick works at Manuel, opened by John G. Stein & Co. in 1930.

The famous Carron Ironworks was established in 1759 and the first foundry in Falkirk 'proper' by the Falkirk Iron Company in 1819. After the Edinburgh & Glasgow, the Stirlingshire Midland Junction and other local railways were opened, around the middle of the century, numerous additional foundries were built, and Falkirk soon became the recognised iron-founding capital of Scotland. A massive range of goods was manufactured. The foundries were mostly served by the Stirlingshire Midland Junction line, and branches leading from it to Dalderse and Bainsford, also by the Camelon Branch, which left the main E & G line at Falkirk High.

Many other industries were served by the Edinburgh & Glasgow railway, including chemical works, quarries and the West Lothian oil industry. Nowadays this is all in the past (a few oil trains between Grangemouth and Bishopbriggs, notwithstanding) and the E & G has reverted to its original *raison d'être* – a high-speed passenger route between cities.

Consistency of Purpose

If patterns of freight traffic on the Edinburgh & Glasgow have undergone total metamorphosis, the same cannot really be said of the passenger service. In 1842 an express train running between Glasgow Queen Street and Edinburgh Haymarket would have stopped at Castlecary, Falkirk and Linlithgow; nowadays it would typically stop at Falkirk and Linlithgow. Castlecary station has long since closed, but the stop there was to provide 'omnibus' connections to and from Stirling, a function now fulfilled by local trains which leave the main line at Greenhill, a short distance away. In the summer of 1848 the fastest service between the cities took an hour and thirty minutes (admittedly in competition with the Caledonian Railway Carstairs route). Ninety-nine years later, in 1947, the fastest services from Edinburgh to Glasgow took an hour and eleven minutes *non-stop*, and by joining the mid-evening all-stations it was still possible to occupy an hour and fifty-one minutes on the journey. Things have improved somewhat since 1947, of course, but it could be argued that improvements over the last 150 years, in terms of speed and frequency of trains, have been gradual rather than dramatic, and that the passenger trains have fulfilled a fairly consistent purpose throughout the era. Businessmen still make use of the trains to attend important

Typical E & G express passenger locomotive of the mid-Victorian era. 2-2-2 locomotive built by Beyer Peacock of Manchester in 1856. ***D. Martin Collection***

meetings in the 'other' city; Falkirk folk still visit relatives in Linlithgow; family groups from Glasgow still like to enjoy a visit to Edinburgh Castle. Even commuters have been commuting for most of the 150 years, as the experience of Lenzie demonstrates.

This comparative stability on the Edinburgh & Glasgow line has existed in an era when there have been massive changes in other forms of transport. People still alive today can remember when the first aeroplanes flew over Central Scotland, or when the first motor buses ran through the streets of Edinburgh or Glasgow, or perhaps when the first local motor car was purchased by a well-off resident of their village. But nobody can remember the first trains on the Edinburgh & Glasgow Railway – not for a very long time now. A first-hand reminiscence of the earliest trains (though not of the actual opening), sent by an elderly man to the *London & North Eastern Railway Magazine* in 1927, was probably one of the last of its kind recorded. There have certainly been some changes over the years – diesel trains, rebuilt stations, colour-light signals – but perhaps the most significant change has been the great extension, in recent years, of station car-park accommodation. Every week-day large numbers of cars cluster around the station buildings at Linlithgow, Polmont, Falkirk High, Lenzie – yes, and even Croy, while owners fulfil their daily toil in Edinburgh or Glasgow. We do not need to lament the demise of long-gone wayside stations if inhabitants of the adjacent villages still make a train-journey each day – preceded, albeit, by a connecting road trip in their own vehicle.

Trains of Title

By no means all of the trains on the Edinburgh & Glasgow line plied an uncomplicated course between the cities. As previously noted, a wide range of destinations could be reached, by branching off E & G metals, from about 1850 onwards. For the next half-century thereafter, the options were considerably expanded, especially following the opening of the Forth Bridge and its connecting lines, in 1890, although by November 1896 the only Glasgow departures making use of the new bridge were six daily express services to Dundee, at 5.50am, 9.00am, 1.20pm, 4.07pm, 5.35pm and 8.10pm.

Trains running to non-standard destinations include four distinctive 'named' services. The 'Fifeshire Coast Express' was inaugurated in 1910 as an express service between Glasgow and Crail. At a later date the train working became more complex, involving the combination of Edinburgh and Glasgow portions at Dalmeny. After World War II the name was revived, in modified form, as the 'Fife Coast Express' and applied to the 4.07pm. summertime-only Glasgow-St. Andrews express. Five articulated coaches from the pre-war 'Silver Jubilee' streamlined service were employed. The service was discontinued after the 1959 season. The post-war service carried a train-name headboard, at least for a time – an LNER-style headboard features in a photograph of Class B1 locomotive No.61402 climbing Cowlairs Incline at the head of the Express, by E.R. Wethersett, which was published in the January 1951 *Railway Magazine*. Whether the earlier 'Fifeshire Coast Express' ever carried a headboard is unknown, but this seems unlikely, since the 'Lothian Coast Express', introduced by the North British Railway in 1914, is usually considered to be the first-ever named train to carry a descriptive headboard, not just in Scotland, but anywhere in Britain. The service ran from Glasgow Queen Street along the whole length of the E & G main line and carried portions for Dunbar, North Berwick and Gullane. The Dunbar portion did not survive for long and the Gullane portion succumbed at the end of the 1932 summer season, when passenger services were withdrawn from that branch. During the summer of 1933 the train ran to North Berwick only, but the 'Lothian Coast Express' was retired for good at the end of that season. During 1929-31 the train vehicles had included a 12-wheel Pullman dining car.

North British Railway 4-4-0 locomotive No 895 *Rob Roy*, with the 'Lothian Coast Express' at Edinburgh Waverley. *Photomatic*

BRITISH RAILWAYS

WITHDRAWAL OF PASSENGER TRAIN SERVICE FROM RATHO STATION

ON AND FROM

MONDAY, 18th JUNE, 1951

The Passenger Train Service will be withdrawn from Ratho station.

Passengers for Ratho will be booked to Edinburgh from which place a bus service operates.

Passenger train parcels and miscellaneous traffic will continue to be dealt with at Ratho Station and a collection and delivery service will be given by motor vehicle connecting with the passenger train service.

Freight train traffic will continue to be dealt with at Ratho Station. A collection and delivery service will be given by motor vehicle for freight train traffic in less than truck loads.

The 'North Briton' express had its genesis in the year 1910 when an existing Leeds-Edinburgh train was extended through to Glasgow Queen Street. A corridor train thereafter fulfilled the 555-mile round trip between Leeds and Glasgow daily. Full restaurant-car facilities were provided, the catering staff serving breakfast, lunch, tea and dinner during the course of the two-way journey. By the early 1950s departure from Leeds was scheduled for 9.05am, with arrival in Edinburgh at 1.32pm, and Glasgow at 2.57pm. After just over an hour the train left again for the south at 4.00pm, with final arrival back in Leeds at 10.02pm. The 'Queen of Scots' Pullman train first ran in 1928, when an existing London-Edinburgh Pullman service was extended through to Glasgow and provided with new all-steel cars. Two trains were necessary, there being morning departures from both London and Glasgow daily. By the early 1950s departure from Kings Cross was at 12 noon, and from Glasgow at 10.50am, allowing a 12 noon departure time from Edinburgh for the south. The train stopped running to Glasgow in 1964, being reduced to a London-Leeds service and renamed the 'White Rose'.

Today & Tomorrow

Having been taken over by the North British Railway in 1865, the Edinburgh & Glasgow line became part of the London & North Eastern Railway system in 1923 and the nationalized British Railways network in 1948. There were few outward changes during the first few years of nationalization, although stations at Ratho and Philpstoun were closed in 1951. Steam power was still universal, although doomed by the plan for *Modernisation and re-equipment of British Railways*, announced by the British Transport Commission in 1955. An early consequence of this plan was the introduction of Swindon-built diesel multiple units on Edinburgh-Glasgow services, with effect

One of the modernisation-plan 'Swindon' diesel trains, photographed at Dullatur in February 1962. ***Douglas Hume***

from 7th January 1957. The train sets were of six cars and were provided with compact buffet facilities. A few steam-hauled services continued to operate alongside the diesels for some years afterwards. Dr. Beeching's plan for *The Reshaping of British Railways*, published in 1963, proposed closure of the stations at Cowlairs, Bishopbriggs, Croy, Dullatur, Castlecary, Bonnybridge High, Manuel and Camelon, although Bishopbriggs and Croy managed to win reprieves, and survive to this day.

Towards the end of the 1960s the limitations of the Swindon (Class 126) sets were becoming obvious – they no longer measured up in terms of either speed or comfort – and consideration was given to finding a suitable replacement. For various reasons it was decided to revert to locomotive power, although to preserve operating flexibility and facilitate rapid 'turn-round' at the termini investigation of the possibility of 'push-pull' working was carried out. After trials with a single Class 37 locomotive at one end of a train, on this basis, it was decided to employ Class 27s in pairs instead, with a power unit at each end of the train linked by cables along the carriages. A full service, on this basis, was introduced in May 1971, with rakes of up-to-date Mark 2 coaches. The journey time between cities was reduced by as much as 12 minutes, coming down from 55 to 43 minutes. It is interesting to recall that when this best-ever service was introduced in 1971 the Second Class day-return fare between Glasgow and Einburgh was just 90p, which compared with an initial Second Class return fare of 60p (i.e. two single tickets at six shillings each) or a First Class fare of 80p, in 1842.

The Class 27/1 and Class 27/2 locomotives employed on this service did very well for a few years, but the demanding schedule eventually began to take its toll and failures became commonplace. During 1979 a move was made to convert a number of Class 47/4 locomotives to a dedicated Class 47/7 fleet for use on the Edinburgh-Glasgow service. This was achieved by fitting push-pull equipment which would operate through the train lighting circuits. Only one locomotive would be employed per train, 'pulling' in one direction (normally Glasgow-Edinburgh) and 'pushing' in the other. In the down direction the driver would sit in the cabin of a 'DBSO' (Driving Brake Second Open) vehicle, ten of which would be modified from conventional Mark 2f Brake Second Open carriages for the purpose. All other vehicles would be state-of-the-art Mark 3 carriages (five per train). By early 1980 four of the new train sets were in use, and conversion of most of the initial fleet of twelve Class 47/7s had been carried out. However, introduction of a full service was delayed for some months by necessary repairs to Falkirk Tunnel, forcing Edinburgh-Glasgow trains to take the longer route via Grahamston. The new timetable was finally inaugurated on 8th December 1980, with 47-49 minute schedules between the cities.

Now we have Class 158 diesel trains, introduced during 1990, on the line. These are the first entirely-new trains to be provided for the Edinburgh-Glasgow route since the modernisation-plan Swindon diesel sets of 1957. It is anticipated that they will form the backbone of the service for many years to come. The fastest trains currently take 50 minutes, with stops at Haymarket, Falkirk High and Linlithgow or Polmont. It seems well within the capability of the new units to reduce this schedule by some minutes.

Notice advertising a public viewing of the new 'Swindon' diesel trains, 1957.
A.A. Maclean Collection

EXHIBITION
of
DIESEL TRAINS
FRIDAY, 4th JANUARY, 1957
Members of the public are invited to inspect one of the DIESEL TRAINS at
GLASGOW
(QUEEN STREET STATION—PLATFORM No. 1)
or at
EDINBURGH
(WAVERLEY STATION—PLATFORM No. 5)
between
12.30 P.M. AND 4.0 P.M.
On FRIDAY, 4th JANUARY
BRITISH RAILWAYS

History from the Train

The Edinburgh & Glasgow Railway passes through some of the most history-laden territory in Scotland – a hub of activity which witnessed most of the significant stages in the country's evolution, whether political or social, economic or industrial. Features of historical interest which can be observed from a Glasgow-Edinburgh train are virtually limitless, and curious travellers are recommended to obtain some of the local history books mentioned in the **Bibliography** (Page 63) for further information. The following are just a few of the historical themes which can be assimilated from the train.

The Draining of Lenzie Loch

About five and a half miles out from Glasgow, just beyond an overbridge which marks the eastern limit of the former Cadder Marshalling Yard, a small, square brick structure, with a concrete 'lid', can be seen in the middle of field on the up side of the railway. This is the top of one of a number of overflow shafts of the drainage tunnel which formerly decanted the entire Loch of Lenzie (also known as the Gadloch) into a nearby quarry, to drain away down a burn, and eventually into the River Kelvin. The draining was carried out a very long time ago – around 1709-1714 – by a Glasgow merchant named William Marshall, the idea being to release land formerly under water for farming purposes. It can be identified with an early phase of the Agricultural Revolution, when wide-ranging improvements in farming practice were introduced. About a century ago the tunnel became blocked and the loch 'came back'. It can be seen on the down side of the line, just opposite the drainage shaft. Nowadays it is cherished as an important amenity by the residents of Lenzie.

The Antonine Wall

The Antonine Wall was built by the Romans around 142-143 AD, over a 40-mile course from Old Kilpatrick on the Clyde to a point on the Forth east of Bo'ness still to be determined satisfactorily by archaeologists. It was made of cut turves, piled in layers, so over the centuries it has tended to flatten, and is nowadays identifiable as a pronounced feature only in a very few places. Fortunately a deep ditch of V-profile was dug in front of it as an extra fortification, and as this has survived in a much more visible way it is possible to pick out the course of the frontier for long stretches at a time. Forts were constructed at approximate 2-mile intervals along the frontier line, and the remains of some of these can easily be identified.

Travelling eastwards from Glasgow the Antonine Frontier first comes near to the railway on its descent from Croy Hill. The hollow of the ditch can be spotted from the train, perhaps quite difficult to discern, although walkers who have already covered

Castlecary Viaduct, as viewed from the site of Castlecary Roman Fort. *ScotRail*

the ground on foot (see **Stepping out from the Stations** section) should have no difficulty. The frontier crosses over to the south side of the railway, at an oblique angle, a few hundred yards east of the terrace of houses (on the down side) at Dullatur. The pronounced V-profile of the ditch can be seen in the undergrowth, within a wooded plantation. As the train crosses Castlecary Viaduct watch out for the next sign of the Romans, the remains of Castlecary Fort, which appear as disturbed ground within a grove of trees, on the up side just beyond the viaduct. The fort area is in fact intersected by the railway, the construction of which doubtless damaged the archaeological record, but the archaeological record was further despoiled by the archaeologists themselves when they left the site uncovered to the ravages of wind and weather, and predations of the curious, after an excavation in 1902. The frontier line passes back over to the north side of the railway nearby.

With careful observation (and preferably the aid of a good Ordnance Survey map) the most important site along the entire Antonine frontier can be viewed from the train. Travelling on the direct Edinburgh & Glasgow route, via Falkirk High, watch out for Greenhill Junction where the Grahamston line diverges on the up side, then the houses at High Bonnybridge about a mile beyond, then a little loch on the up side, immediately beyond that again. Just past the loch a grove of trees comes into view, with an embankment running within it, almost parallel to the railway. This embankment is the best-preserved stretch of the turf rampart of the Antonine Wall, clearly surviving here (to a height of about 5 ft) for some distance. After about half-a-mile an earthen mound, surrounded by deep ditches, can be discerned. This is the remains of Rough Castle – best preserved of all the Antonine Wall forts. Until 1967 it was possible to visit the fort by taking the train to Bonnybridge High Station – nowadays it is necessary to proceed to Falkirk and catch a Bonnybridge bus.

Falkirk Tryst

No clear feature marking the site of Falkirk Tryst can be seen from the train, but the episode in Scottish history was so important that it merits a thought while passing the appropriate locality. The Trysts were massive livestock markets held on the second Tuesdays of August, September and October each year, when huge numbers of cattle, sheep and horses changed hands. Drovers arrived from all parts of Scotland, but especially perhaps the Highlands, and their animals were sold, often as not to dealers from south of the Border. Until the 1770s the Tryst was held at Redding Muir (a locality to the south of the railway, between Falkirk and Polmont). Then it was moved to the Rough Castle/High Bonnybridge area, also quite close to the E & G railway route, where, however it was hampered by old coal and iron workings, and the presence of the Forth & Clyde Canal. Finally, after 1785, it was transferred to Stenhousemuir, well to the north of the railway route, on the other side of the Carron Valley. The coming of the railway can be blamed for the decline of the Tryst, since it made the drovers' overland trek unnecessary. Dealers could travel by train to local markets in any part of the country, and purchased livestock could then be conveyed to the required destination in rail-borne cattle-trucks.

The Windmill at Myrehead

Windmills are not particularly common in Scotland, and none survive with their sails intact, so it is gratifying that a reasonable example can be observed close to the Edinburgh and Glasgow railway line. This is at Myrehead Farm, on the up side of the line two miles east of Polmont, quite close to the site of the former Manuel Station (where the embankment of the former low level line can still be seen at right angles to the main line). This example is of the 'tower windmill' variety, and would formerly have had a revolving cap, which like the sails is no longer present. The absence of the cap gives it a 'cut-off' appearance. It is believed to be about two hundred years old. A short distance to the east, at another farm, also on the up side of the line, can be seen a low, wide circular building, attached to a farm building. This is an example of a horse-gin mill – the more common alternative to a windmill. Instead of wind-power the threshing-mill employed the services of sturdy farm horses, walking round and round in circles within the distinctively-shaped building. Until about twenty years ago an open-air horse-mill could be seen quite close to the railway, at Loch Farm beside Lenzie Loch, with the wooden beam, to which a single horse was formerly yoked, exposed.

Also close to the windmill at Myrehead is the northern extremity of the preserved railway line owned by the Scottish Railway Preservation Society, which extends from here to Bo'ness.

Linlithgow Palace

Some of the features mentioned in this chapter might seem routine and mundane – but not so Linlithgow Palace, birthplace of Mary Queen of Scots and one of the key historical structures in Central Scotland. The Palace is sited to the north of the railway and the view of it from the railway is masked, to some extent, by the adjacent St. Michael's Parish Church, with its startling wooden spire, dating from 1964. Work on the present Palace was begun in 1425, following a destructive fire which swept through the town the previous year, destroying earlier buildings on the site. It was a very long time building, and various kings of Scotland added personal touches of their own. James V, for example, was in 1538 responsible for the elaborate courtyard fountain, which seemingly was made to flow with wine on occasion. Just four years later Mary Queen of Scots was born in the Palace, so guaranteeing an additional mention for Linlithgow in the history books. After James VI & I left Scotland to accept the unified throne in London, in 1603, the palace fell into neglect and the structure deteriorated. The roof of the north wing collapsed in September 1607. This brought about the last great construction project to be undertaken at the Palace, the rebuilding of the north range in fine Renaissance style, which was carried out during 1618-24, towards the end of the reign of James VI. The circumstances which caused the ruination of Linlithgow Palace were sad in the extreme. On 31st January 1746 some of the Duke of Cumberland's troops in pursuit of Bonnie Prince Charlie spent the night in the palace. When they left the next morning they did not trouble themselves to put their fires out, and the flames shortly ignited the building. To the carelessness of the English troops was then added the indifference of local townspeople, and the palace was burnt out. Nobody lifted a finger to save it. All the splendours went up in smoke and it has remained a ruin ever since.

The Binns Tower

High up on a ridge to the north of the railway, just over two miles east of Linlithgow is a cylindrical stone tower with a crenellated top, surrounded at the base by a circular stone wall. Although a distinctive landmark it serves no useful purpose and is usually described as a 'folly'. It was erected by James Dalyell in 1826, with money won in a wager. Situated nearby is the House of the Binns, ancient seat of the famous Dalyell family. The present mansion was constructed during the 1620s by Thomas Dalyell, who had made his fortune in London as a member of the entourage of King James VI & I. In 1642 Thomas Dalyell was succeeded by his son, the famous Royalist soldier General Tam Dalyell, who defeated the Covenanters at Rullion Green in 1666 and raised the Scots Greys regiment in 1681. The House of the Binns was gifted by the Dalyell family to the National Trust in 1944, but the present Tam Dalyell, the well-known local Member of Parliament, still has apartments there.

Niddry Castle

A short distance from the eastern portal of Winchburgh Tunnel, on the north side of the railway, is the picturesque Niddry Castle. This is a four-storey tower house built on an L-shaped plan, in the late fifteenth or early sixteenth century. For many years it was no more than a roofless, dilapidated ruin, but in 1984 it was purchased from the National Trust for Scotland by a community education worker with Lothian Regional Council, for £7,100. In a comparatively short space of time he transformed the ruin into a comfortable home, forcefully demonstrating the potential of the many similar mediaeval ruins in Scotland. In July 1991 it was put up for sale, with offers over £270,000 invited. It was described as 'An Intriguing Restoration of a Traditional 15th Century Tower House' with 'Original Great Hall, Guard's Room, Gallery/Store. Two Fine Reception Rooms, Kitchen/Breakfastroom, Master Bedroom Suite, Four Further

Niddry Castle, Winchburgh, showing part of the adjacent shale bing, to the left. ***ScotRail***

Bedrooms, Second Bathroom. Usual offices. Gas Central Heating.' According to a local tradition Mary Queen of Scots spent her first night here following her flight from imprisonment at Loch Leven Castle in 1568. It deserves to be true.

The Shale Bings

Rising behind Niddry Castle, and completely dwarfing it, is an enormous shale bing, one of several in this area. They stand as monuments to the West Lothian shale oil industry, which was of vital importance to the local economy throughout the second half of the nineteenth century and well into the twentieth. The local oil industry had its origins in the 1850s, following James 'Paraffin' Young's 1850 patent to distil oil from cannel coal or 'torbanite'. With limited supplies of this mineral available, the future of the local industry was secured by the discovery of oil-shale at Broxburn in 1858. The huge bings of spent shale which accumulated are typically flat-topped, of a shape quite distinctive from normal coal bings. A branch line to Oakbank Oil Works left the main line on the up side, near Niddry Castle. A short distance away, on the down side, the Broxburn branch curved away, to serve the Albion Oil Works and Broxburn Oil Works.

South Gyle

Approaching Edinburgh from the west, the first scattering of buildings to be encountered on the up side is the recent commercial development at South Gyle, with numerous warehouses and factories. A railway station was opened to serve the locality on 9th May 1985, but on the line which curves away northwards to the Forth Bridge, not the Edinburgh and Glasgow main line. As well as providing a rail-service for this developing community, the £130,000 station was intended as a 'good railhead' where commuters from a wider area could park their cars before proceeding to Edinburgh, and thus avoid the road congestion which exists on the western approach to the city. Spaces for 60 cars were provided.

An interesting feature to be seen at South Gyle, on the up side of the line, is a Lightning jet fighter plane, mounted on a large pedestal. It was presented by British Aerospace to G.E.C. Ferranti to symbolize the teamwork and engineering skills 'which provided the world's first monopulse radar for the Royal Air Force's first supersonic fighter'. A plaque was unveiled by Mr R.H. Evans, Chief Executive, British Aerospace plc, on 15th June 1990. The Lightning can be observed just beyond a large red-roofed pagoda-like building which forms part of the G.E.C. Ferranti complex at South Gyle.

Murrayfield

Coming closer to Edinburgh, Murrayfield Stadium is very obvious, right beside the railway on the up side. By the early 1920s the Scottish Rugby Union's former international ground at Inverleith was unable to cope with the increased attendances then being attracted to matches. In 1922 a polo ground at Murrayfield was purchased for redevelopment as a rugby stadium. The new ground was ready in time for the Scotland v England match on 21st March 1925, which in the event attracted a crowd of 70,000, more than double the record attendance at Inverleith. Scotland won 14-11, and in so doing achieved the so-called 'Grand Slam', the 'Triple Crown' and the Calcutta Cup, a feat not equalled by Scotland until season 1983-84, when the task was completed

'Lightning' jet fighter plane at South Gyle. ***ScotRail***

with a memorable 21-12 win over France at the Stadium, on 17th March 1984. Other notable games at Murrayfield have been a 13-7 victory over England on 17th March 1990, when Scotland's third 'Grand Slam' was achieved, and the semi-final of the World Cup on 26th October 1991, when England defeated Scotland 9-6. Also worth a mention is the Scotland v Wales match held on 1st March 1975, which was attended, very uncomfortably, by a record crowd of 104,000. As a result of this, 'all-ticket' matches became the order of the day. Many improvements have been made to the ground over the years, including, in 1981, provision of a new East Stand. A significant innovation in 1959 was an undersoil heating system. Further improvements are planned, for the near future.

Princes Street Gardens

Between Haymarket and Waverley stations, in Edinburgh, all trains pass through the famous Princes Street Gardens. The origins of the gardens date back to the year 1816, when an Act of Parliament was passed, effectively prohibiting the construction of buildings on the south side of Princes Street, and authorising the proprietors of houses on the north side to contract with the Corporation for a lease or feu of the ground on the opposite side 'for the purpose of laying out the same in whole or in part in a garden, nursery for trees, or pleasure ground or under grass or otherwise embellishing and enclosing the same' at the expense of the Princes Street Proprietors. A Committee of the latter was appointed on 24th June 1816, for carrying this into effect, and thereafter the amenity of the gardens was jealously guarded. In 1837 attempts by the Edinburgh & Glasgow Railway to obtain sanction to take their line through were successfully opposed, and the E & G settled for a terminus at Haymarket at that time. However, with the prospect of the North British Railway's link to England reaching the eastern end of the gardens the E & G tried again, this time with success. To pacify

Murrayfield Stadium, as viewed from the railway. ***ScotRail***

Train in Princes Street Gardens. ***D. Martin***

the Proprietors it was agreed that W. H. Playfair, architect of the Royal Institution on the Mound, should undertake to enclose the railway by a rubble stone wall on each side, 6 feet high, and raise a mound of earth along the north side, above the level of the lines, to conceal the railway from the drawing-room windows of houses in Princes Street, the mound to be sloped down gently towards the garden level, with which it was to be gracefully united, and 'by a profuse use of ivy, evergreens and trees the presence of the railway would become undiscoverable, and the usual appearance of the gardens would be preserved'.

Trains first started to run through the Gardens in 1846, and they have carried on doing so ever since, shielded at all times by Playfair's walls, mound and evergreens or their successors. Nevertheless, there has been a great deal of continuing unhappiness concerning the presence of the trains in the Gardens. Even some railway enthusiasts have seen the other point of view. Canon Roger Lloyd, in a hedonistic little book on *The Fascination of Railways* (1951) pronounced the line through Princes Street Gardens to be:

> ...a black offence, since they shunt in the the very gardens, and the stationary engines, waiting to take over the trains arriving at Waverley station, pour out black smoke, and their safety valves scream all night through.

However, the spectacle created by trains scurrying through in a constant succession can itself be seen as an amenity. Generations of parents have treated young children to a picnic in the gardens, in the knowledge that a trip to one of the adjacent railway footbridges would be a real highlight of the day.

During the summer of 1990 plans were laid to introduce overhead electric wires to the railway lines through the gardens, in connection with the extension of the West Coast electrified network from Carstairs to Edinburgh Waverley. It was clear that the two remaining iron bridges of the old-fashioned type, with iron lattice-work, would need to be taken out of use, since they offered insufficient protection from the dangerous electric wires. In November a disappointed parent, Morna Munro of Bruntsfield Place, Edinburgh, wrote to the *Scotsman*, expressing dismay 'at the total disappearance of the lovely old wrought iron pedestrian bridge behind the Ross Bandstand in Princes Street Gardens' where she had spent many happy hours with her daughter, waving at trains. It transpired that all was well, and a new wooden bridge with special toughened glass 'viewing panels', to enable children to see the trains, was on its way. It was officially opened by Miss Flora Munro, aged 6, on 22nd January 1991.

Stepping out from the Stations

A Walk round Lenzie

Lenzie is an interesting place because its development and growth was influenced by the Edinburgh & Glasgow Railway in a quite intimate way. Before the railway came there was nothing here but a few farms, one or two landowners' houses, some weavers' cottages, and a damp and dreary moss. The original station of 1842 (known then as 'Kirkintilloch') was on the present site, but was moved some distance to the east in 1844 to facilitate passenger exchange with the 4ft. 6in. gauge Monkland & Kirkintilloch Railway, before being moved back in 1848 to coincide with the opening of the Campsie Branch Railway. The first dwelling houses of the modern community can be dated precisely to this latter move of 1848, and the development of the place as a residential locality can be mapped very precisely from that period onwards. Housing development was given a boost by the Edinburgh & Glasgow Railway Company's offer, dating from about 1850, to provide free tickets to persons building substantial villas close to any of their stations, although real growth had to wait until the 1870s, when a supply of running water (an essential amenity) became available for the first time. The railway company also influenced the name of the community, which was known as 'Campsie Junction' in line with a station renaming of 1849, and became 'Lenzie' when the station was renamed 'Lenzie Junction' in 1867.

Leave LENZIE STATION by the south platform exit and turn left down the lane which runs parallel to the railway. On reaching the road at the foot of the lane, notice the gate on the corner at the right-hand side. This belongs to one of the first three railway-influenced houses of 1848 – the other two lying immediately to the south, in line along the road. Turn left under the railway and continue along the road for a short distance until the station car-park comes into view on the left-hand side. The bend to the left immediately north of the railway was much sharper here until a realignment of 1963 (when, also, the present underbridge was built) and the road was much narrower; until this date a mirror was positioned just north of the railway, so that traffic coming in the opposite direction could be seen, and collisions avoided. Just past the car-park, on the left-hand side, is a row of shops. Notice that in fact these are made up of two groups – a high tenement block, followed by a lower, two-storey building. This latter building, dating from about 1865, contains the first shops to be built in Lenzie after the coming of the railway. Straight across the road is Lenzie Old Parish Church – one of the three principal churches in Lenzie (all of which date from the 1870s, an indication of the rapid growth which took place locally during that decade). The Old Parish Church is situated on the south corner of Garngaber Avenue; on the north corner notice an impressive, castellated villa, known as 'The Tower'. This building dates from 1858 and was one of these benefitting from the Edinburgh & Glasgow Railway's offer of free tickets 'to parties who may build villas within a mile of any of the Stations on the line'. As might be expected, all of the Lenzie houses in this category lie close to the railway.

Continue northwards over the brow of the hill along the main road to Kirkintilloch (known here as 'Kirkintilloch Road'). A short distance downhill, on the left, is Lenzie Primary School. Although totally rebuilt, this school is on the site of the original 'Lenzie Academy', a private school of 1870. In 1886 it was replaced by a Board School of the same name, which lasted on this site until the construction of a modern school elsewhere (in Myrtle Avenue) in 1960. Just beyond the primary school, up a short lane, is a villa known as 'Pomona', former residence of the first Lenzie Academy rector of the 1870s. Across the road from the primary school note Lenzie Union Parish Church, another of the churches built in the 1870s. A short distance further along turn left into Hawthorn Avenue. Beyond a small group of modern houses, the second stone villa on the left, known as 'Netherby', was for many years the dwelling of Matthew Holmes, locomotive superintendent of the North British Railway, who designed and built a great many engines, some of which outlasted his death in 1903 by more than half a century. One of them – the 0-6-0 No.673 *Maude* – is preserved by the Scottish Railway Preservation Society at Bo'ness.

Return to Kirkintilloch Road and continue northwards. The next avenue on the left is Viewfield Avenue, which until recently consisted entirely of houses of the early growth period of the 1850s – notice 'Willowfield' right on the corner and two unusual semi-detached or 'double villas' a short distance along. The next avenue again on the left, is Myrtle Avenue, followed by Beech Road. Notice the small villa between these two – this was originally 'Auchindale', built during the 1850s by J.B. Thomson, Assistant General Manager of the Edinburgh & Glasgow Railway. He is said to have been a most influential figure in the early development of Lenzie. According to local tradition he had an early telegraph wire laid to his house, so that he could provide menu instructions for his evening meal! Turn left into Beech Road, and notice the second large villa on the left. This is 'Norwood', one-time residence of Dugald Drummond, famous locomotive superintendent of, successively, the North British Railway, the Caledonian Railway, and the London & South Western Railway. A short distance further along Beech Road, on the right-hand side, is St. Cyprian's Episcopal Church, third and last of the Lenzie churches of the 1870s (actually the first built). Continue to the end of Beech Road, then turn right into Gallowhill Road. Continue along this street for some distance, past Marguerite Avenue and Gallowhill Avenue on the right, until the main Lenzie-Kirkintilloch Road is re-joined at a location known as 'Whitegates'. This place was named on account of the level-crossing gates of the ancient Monkland & Kirkintilloch Railway, which crossed the road here until final closure in 1966. The gates have now gone, but notice a plaque on the far wall, a short distance north of the road junction, commemorating the former presence of this historic railway.

The plaque marks the starting-point of a walk-way along the Monkland & Kirkintilloch trackbed. The M & K was opened in 1826 to convey coal from the Monklands to the Forth & Clyde Canal at Kirkintilloch – most of it destined for the Edinburgh market. It was the second public railway (after the Stockton & Darlington) to be laid with the excellent Birkinshaw patent rails, so was very able to support locomotive traction – the use of engines in fact began in 1831 with the first-ever pair of Glasgow-built examples. Turn southwards along the walk-way and continue for about half- a-mile until another pathway converges on the left – this is the trackbed of

'The Tower', Lenzie – one-time 'Villa Ticket' house. ***Strathkelvin District Libraries***

'Norwood', Beech Road, Lenzie – one-time home of Dugald Drummond, the famous locomotive engineer. ***Strathkelvin District Libraries***

a link-line to the Campsie Branch Railway of 1848. The junction here was known as 'Woodleys Junction'. Continue along the 'main-line' until the site of an obvious level-crossing is reached. Turn right into the road here and continue along Garngaber Avenue back to LENZIE STATION. As 'The Tower', previously mentioned, is approached on the right, note the three old villas nearest to it on the same side of Garngaber Avenue. (A modern house has recently been inserted in among them). These three villas, like 'The Tower', are 'villa ticket houses', dating from 1856-8. One of them, the house now known as 'Hillside', was also, at one time, the residence of Thomas Wheatley, another locomotive superintendent of the North British Railway. Among other claims to fame, Wheatley designed No.224, the engine which fell with the First Tay Bridge in 1879. (WALKING TIME, LENZIE STATION TO LENZIE STATION, VIA KIRKINTILLOCH ROAD, BEECH ROAD, WHITEGATES & MONKLAND & KIRKINTILLOCH WALK-WAY: APPROXIMATELY 1½ HOURS).

From Croy to the Antonine Wall

The Antonine Wall was built by the Romans about 142-143AD to further contain wild northern tribes previously stemmed by Hadrian's Wall in the north of England. Unlike Hadrian's Wall which was built of stone, however, the Antonine Wall was made of cut turves piled in layers, so it has not survived nearly as well. Fortunately, though, the Romans dug a deep ditch of V-profile in front of the wall, as an extra fortification, and long stretches of this survive in good condition. Clear remains of some of the forts, which were built at intervals along the wall, can also be seen. Like the Forth & Clyde Canal, the Antonine Wall runs close to the Edinburgh & Glasgow Railway for a considerable distance. From about Bishopbriggs to some way east of Falkirk, indeed, it is never more than a mile and a half from the railway. It crosses the main line east of Dullatur and again near Castlecary; it also crosses the Grahamston loop near Bonnybridge and again just east of Falkirk. Remains of two of the forts, at Castlecary and Rough Castle, can clearly be seen from the train. The only stretch of the wall to survive in reasonable condition, at Rough Castle, can also be seen from the train, as can the ditch at several locations. A number of railway-based walks can readily be devised to explore sections of the frontier on foot. Croy is probably the most convenient starting point.

Walk One. Ascend to the overbridge at CROY STATION and walk northwards along the entire length of the village main street (Constarry Road). After leaving the houses behind, turn right at a road junction (towards Kilsyth), and a short distance along notice a sign 'ANTONINE WALL BAR HILL ½ MILE' on the left-hand side of the road. Take the vehicle-width track here, in the direction signposted, and continue along for about a quarter of a mile until a metal gate, with side wicket, leads into a plantation of trees. A short distance further on a wide clearing is reached. Turn sharp right off the track here, over a very pronounced ridge. This is the ridge of the Antonine Wall, and beyond it the bracken-filled depression of the V-profile ditch is encountered. It can be followed westward for a distance, soon climbing up onto high ground at Castle Hill. Just below the summit of this hill an excellent view of the frontier line is obtained by looking back eastwards – beyond the stretch we have just followed the line of the ditch can be seen winding over the rocky crags of Croy Hill and disappearing

View looking east from Castle Hill, Twechar, with the v-profile of the Antonine ditch in foreground, and its course over Croy Hill visible in the distance. ***Strathkelvin District Libraries***

into the distance. This serves as a reminder that the Antonine frontier can be followed eastwards from here most of the way to Falkirk, with just a few short breaks where it has been obliterated. An excellent guide to following its course (described in an east-west direction) is Professor Anne Robertson's book *The Antonine Wall*, the fourth edition of which, revised and edited by Lawrence Keppie, was published by the Glasgow Archaeological Society in 1990. From the summit of Castle Hill, at the triangulation pillar (about 500ft. above sea level), a panoramic view both east and west is obtained, with, to the north, the Campsie Fells and Kilsyth Hills sweeping right across the line of vision. On a really clear day a glimpse of both the Forth and the Clyde can be obtained, reminding us that we are here very close to the centre-point of the narrow, 40-mile waistband in the centre of Scotland. To the west the sweep of the Antonine ditch can be seen in the foreground, and as it disappears from view its line is taken up by the Forth & Clyde Canal beyond, a clear indication that these two man-made constructions across the isthmus are never very far away from each other.

From the summit of Castle Hill a path winds down in a south-westerly direction to a grove of trees on a wide plateau. This is Bar Hill, site of one of the most interesting forts on the Antonine line. Half-an-hour or so should be allowed to examine its remains. First find the well, which is situated right in the centre of the fort and has been stoutly consolidated, re-using stones from the original Roman construction. The well has an interesting archaeological history, dating back to an excavation carried out

1902-5 by the prominent archaeologist George Macdonald, with the patronage of Alexander Whitelaw, the local landowner. On the very first day of the 3-year project, 20th November 1902, a decision was taken to examine a particularly green patch of grass, and very soon the stonework of the well was encountered, just underground. As it was painstakingly cleared of debris, it became clear that this was a 'treasure trove' of some significance, for the Romans had deposited numerous artefacts in the well during an apparently hasty retreat – an altar; fragments of a building inscription; stone columns with their bases and capitals; building stones; miscellaneous wooden objects; the well-bucket; leather items, including a few shoes; iron weapons and tools; bronze ornaments; coins. This great discovery, achieved in such a dramatic way, undoubtedly gave a great initial boost to twentieth century Roman archaeology in Scotland. Around the well can be seen the foundations of the headquarters building of the fort; immediately to the south is the site of an assembly hall for the fort garrison, with a dais at one end for use by the commander while addressing his troops; behind are the remains of the fort offices, including a chapel where standards would have been kept on display. After examining the fort H.Q. the next thing to look for is the bath-house in the north-west corner of the site. A shaped stone at the eastern end indicates the position of the furnace, and in line from here are the remains of the various apartments, including the *caldarium* (hot room, closest to the furnace), the *tepidarium* (warm room) and the *frigidarium* (cold room). In the hotter apartments can be seen stone hypocaust pillars, the remains of an underfloor heating system.

Leave the fort vicinity by a gate in the south-west corner and pass westwards along a short path before turning right into a rough farm road. This winds downhill for some distance, until joining a public road at Twechar War Memorial. Turn right along this road for a short distance until the Forth & Clyde Canal is encountered at Twechar Bridge. Notice the abutments of a previous bridge alongside to the left – these belonged to the old wooden bascule bridge, which surprisingly was replaced by a swing bridge as late as 1960, only two years before the canal was closed to through navigation. The swing bridge is still in position but hasn't 'swung' since the end of 1962. Cross this bridge and turn left along the Kirkintilloch road. A short distance along notice the pivot of another former swing bridge across the canal, close to the south bank. This was for a colliery railway, owned latterly by the National Coal Board. The bridge was provided during the 1860s for the use of William Baird & Co. on condition that they would 'maintain the quantity of coal traffic that they sent by canal'. This led to a situation, in the early twentieth century, whereby Bairds were still sending coal along the canal at a time when other firms had transferred it all to the railways. Close to the bridge on the south bank was a substantial N.C.B. works and locomotive depot, all traces of which have disappeared. Both works and bridge were closed in 1963.

Return to Twechar Bridge and continue eastwards along the north bank of the canal for about a mile and a half to the next bridge, at Auchinstarry. Turn right here and continue along the B802 road, back to CROY STATION. (WALKING TIME, CROY STATION TO CROY STATION, VIA BAR HILL, TWECHAR & AUCHINSTARRY: APPROXIMATELY 3 HOURS).

Walk Two. Proceed northwards from CROY STATION through Croy Village as in Walk One, but shortly before the Kilsyth road junction turn right into Nethercroy Road. A sign here reads 'ANTONINE WALL CROY HILL 400 YARDS'. Almost immediately a track swings round to the right, to begin the ascent of Croy Hill. Carry on past a small disused quarry, on the left, then turn left at the corner of some houses and head straight for the summit of the Hill. The Antonine ditch now appears very obviously on the left and continues in impressive fashion for some distance, in places hewn out of the solid rocky crags. The path runs along to the south of the ditch, gradually ascending to the summit of Croy Hill. From here there is a really excellent view of the Kelvin Valley, sweeping across the line of vision from north-west to north-east, with the Canal far below and the Campsie Fells and Kilsyth Hills beyond. The town of Kilsyth also figures prominently. Looking to the south-west, a clear view of the course of the Edinburgh & Glasgow Railway is obtained, as it proceeds westwards from Croy Station across a small viaduct.

On the eastern descent of Croy Hill the line of the Antonine ditch becomes very obvious once again, before disappearing round the brow of another, smaller, hill. The path ascends to the summit of this hill, and on the far side the ditch can be seen yet again, following a tortuous course down below. The site of the Antonine Wall fort of Croy Hill now comes into view, marked by a group of six distinguished-looking trees. The fort has been excavated on a number of occasions, although little evidence of this can be seen. Croy Hill was apparently one of the smaller forts, with an area of only 1.5 acres, but evidence of a headquarters building, a granary and a well has been found, also a bath-house (outside the main precinct). The fort is 400ft. above sea level.

Just beyond the site of Croy Hill fort is a surprising feature, for here a 25yd. stretch of the Antonine ditch was seemingly never dug out by the Romans. It seems strange in the light of the arduous work involved in hewing the ditch out of solid rock on the slopes of Croy Hill that this short stretch should have been neglected, but it was so. The purpose (or excuse!) will very likely remain a mystery for ever. Beyond this 'missing bit' the ditch resumes its impressive profile and continues eastwards in style for some distance. The path follows on, some way to the south of the ditch, and soon Dullatur village can be seen in the distance, over to the south-east, with the Edinburgh & Glasgow railway skirting it on the north side. There was formerly a passenger station here, but it closed in 1967, under the Beeching plan.

A short distance further along, the ditch and path encounter the embankment of a former mineral railway, crossing at right angles. Leave the line of the ditch here, and turn left along the embankment, at the same time noting that the V-profile depression continues onwards, soon leaving the rough terrain and forming a gentler hollow across a grassy field. Very soon the Forth & Clyde Canal comes into view once again, with the Dullatur-Kilsyth road in between. The road is reached by means of a track, branching to the right away from the railway embankment. By turning to the left along the road for a short distance, thereafter, the Forth & Clyde Canal is presently crossed at Craigmarloch Bridge.

Craigmarloch is a most interesting location, with a special place in the hearts of elderly people from Glasgow, Bishopbriggs and Kirkintilloch. It was the destination of James Aitken's famous fleet of pleasure steamers on the Canal, the *Fairy Queens* (three of them, at different times), the *May Queen* and the *Gipsy Queen*. Fond memories are retained of picnics here, or of teas in the Bungalow Tea-room, which stood just east of the bridge, on the south bank. Other facilities included swings, small boats and a putting green. Also at this location, the important supply of water from Townhead Reservoir (known to the locals as Banton Loch) enters the canal from the north, by means of a water-course or lade. A path runs alongside this feature to the ruins of the former canal stables, some distance back from the towpath. Now roofless, the stables are still nonetheless impressive, and it is to be hoped that a useful purpose will soon be found for them, as with similar structures elsewhere on the Canal. These were canal company stables, of a standard pattern, and were at one time occupied by horses for the 'Swift' service of fast boats, introduced during the 1830s in an attempt to counter the threat from rail transport. Craigmarloch Bridge is modern, but interesting, being one of the recent bridges built to the new '3-metre headroom' rule, to permit navigation along the Canal by pleasure craft.

Turn westwards along the towpath (on the north bank) for some distance. Just before Auchinstarry Bridge, the Forth & Clyde Canal Society's trip boat *Gipsy Princess* can often be seen in a small basin, opening off the south bank. The *Gipsy Princess*, introduced in 1990, is 45ft. long and is licensed to carry 36 passengers. She provides public cruises on summer Sundays and is also available for hire. Her base is reached by a track running along the south bank from Auchinstarry Bridge. A former canal stables building survives on the other side of the bridge, also on the south bank. It was formerly attached to the old Auchinstarry Inn, built to serve the needs of canal passengers, and demolished only in fairly recent years. The stables block is now used as a cowshed.

Return to CROY STATION by way of the B802 road, running south from Auchinstarry Bridge. (WALKING TIME, CROY STATION TO CROY STATION, VIA CROY HILL, CRAIGMARLOCH & AUCHINSTARRY: APPROXIMATELY 2½ HOURS).

Note. The two Antonine Wall walks from Croy Station can be combined, either by continuing along the canal bank eastwards from Auchinstarry Bridge, towards the end of Walk One and thereafter accomplishing the Croy Hill walk in reverse order; or by continuing westwards along the canal bank from Auchinstarry Bridge, towards the end of Walk Two, and thereafter visiting Twechar and Bar Hill in a direction opposite to that described. (WALKING TIME, CROY STATION TO CROY STATION, VIA BAR HILL, TWECHAR, AUCHINSTARRY, CRAIGMARLOCH & CROY HILL: APPROXIMATELY 5 HOURS).

Maps of the Bar Hill (slightly extended) and Croy Hill walks are published in a set of three 'walk cards' of the Antonine Wall/Forth & Clyde Canal vicinity, available from British Waterways, Rosebank House, Main Street, Camelon, Falkirk, FK1 4DS, at a cost (1992) of £2.75, post free.

Falkirk Meandering

With its uniquely central position in Scotland, and its impressive-looking parish church and town steeple, Falkirk seems to beckon for a closer inspection of its credentials as an ancient burgh. To some extent, however, close inspection yields disappointment. The parish church is a rebuilding-job, dating largely from 1810-11; the town steeple is a replacement, dating only from 1813-14; the cross well (absent for repair at the time of writing) is a replica of the original, dating from 1817; the mercat cross itself is nowhere to be seen, although its former position is marked by a cross-pattern in the cobbles, close to the town steeple; and, most frustratingly of all, the expected proportion of crow-stepped houses with steeply-pitched roofs, customary in the older Scottish burghs, is totally absent. Yet Falkirk has undoubted character – when walking along the pedestrianised High Street the spirit of the old burgh seems present in the air. Everything on view might indeed be replica, replacement or rebuilding,

Tomb of Sir John de Graeme, at Falkirk Parish Churchyard. *Strathkelvin District Libraries*

but at least the 'goodwill of the business' survives in ample measure. A pleasing feature is the number of interesting old street names which have survived: Baxter's Wynd, Cow Wynd, Kirk Wynd, Wooer Street, Howgate, Lint Riggs – even 'Glebe Street', possibly the most evocative of all Scottish street names.

Elsewhere in Falkirk, a most important survival is the infrastructure of the Forth & Clyde and Union canals. The canals were vital to the economic well-being of the burgh for a long period of time, so it is quite appropriate that canal fixtures and buildings should form a significant proportion of Falkirk's surviving historical legacy.

To explore Falkirk to best advantage, leave FALKIRK HIGH STATION by the main north-platform entrance, skirting the eastern part of the car park, and proceed north-eastwards into High Station Road. High Station Road soon leads into Cow Wynd, so called, it seems, because cattle were driven along it to common pasture in the vicinity of the modern Booth Place (a street adjacent to Cow Wynd on the left-hand side). At one period the Town Council decided to apply the name 'High Station Road' to the whole of the High Station Road/Cow Wynd thoroughfare, but found that it was one thing to take a bureaucratic decision and quite another to have it popularly adopted – so 'Cow Wynd' remained 'Cow Wynd' and does so to this day. At the end of Cow Wynd, turn left into the pedestrianised High Street, and notice immediately the imposing town steeple, designed by David Hamilton, the well-known Glasgow architect and erected during 1813-14. Nearby is the cross well, and alongside is Tolbooth Street, formerly registered in the *Guinness Book of Records* as 'the shortest street in Britain' (58 feet, or 17.67 metres). At the base of the steeple will be found the Falkirk

Information Centre, where advice and literature can be obtained to help with further exploration of the area. Proceed westwards along High Street until the distinctive Parish Church comes into view on the right-hand side. It is well worth while lingering here to examine the church and its environs. Near the south-east corner of the churchyard is the tomb of Sir John de Graeme, who was killed while fighting in support of William Wallace against the English army at the first Battle of Falkirk in 1298. The monument is a complex structure, consisting of a recumbant stone effigy, surmounted by three horizontal stone slabs and enclosed within a decorative iron cage. The different components are of varying age, representing development and evolution over an extended period, perhaps four hundred years. Nearby is a simple stone slab, according to the (fairly modern) inscription upon it the grave of Sir John Stewart, another Battle of Falkirk hero. Nearly seven hundred years have elapsed since the first Battle of Falkirk, and it is now impossible to prove whether Sir John de Graeme and/or Sir John Stewart are buried under their respective stanes. What is important, however, is that numerous generations of Falkirk natives have *believed* that they were, and have cherished their memory and cared for their supposed relics. Together they form a significant strand of local folklore and culture. This being the case, it is unfortunate that the Sir John Stewart stone lies perilously close to a vehicular access road, and displays signs of recent wheel damage.

Leave the churchyard by the east gate, situated close to the memorials to the two Battle of Falkirk heroes, and pass along the short Manse Place. Turn left into Vicar Street and descend to the Newmarket Street crossing, noting on the left the equestrian statue of the Duke of Wellington, which formerly stood in High Street, adjacent to the steeple, but was moved to this location in 1905. Pass straight over and continue along Vicar Street in a northwards direction. Beyond a pedestrian underpass FALKIRK GRAHAMSTON STATION comes into view on the left-hand side – regular services to this station can be used if a brief inspection of Falkirk town centre is sought. Continue straight on into Grahams Road. Some way along it is worth sidetracking left into Gowan Avenue, for a short distance, to inspect the large cast-iron gateway of the Grahamston Iron Company's foundry, one of the few substantial monuments still proclaiming Falkirk's heritage as a major iron-founding centre. Return to Grahams Road and continue to the Forth & Clyde Canal crossing at Bainsford.

The section of canal through Falkirk was part of the first stretch to be opened, from the Forth to Kirkintilloch, in 1773. Completion through to the Clyde at Bowling was finally achieved in 1790. A set of twenty locks in the Falkirk area raised the canal up through 156ft. from sea level at Grangemouth to a long level stretch across the middle of the country. Another set of twenty locks in the west took the canal back down to the Clyde. The remains of Lock No.5 can be seen along to the right as we cross the canal at Bainsford. The opening of the canal brought immense economic benefit to the towns and villages on its banks, which previously had to rely on poor roads for the transportation of goods. A range of canal-side industries began to develop, notably iron-founding at Falkirk. The wooden bascule bridge at Bainsford was one of the first standard Forth & Clyde Canal bridges to disappear, when the arrival of electric tramcars in the area in 1905 necessitated the construction of a substantial swing bridge at this point. Unfortunately the uprights supporting the tramwires obstructed the towpath, so the introduction of trams across the bridge was delayed for some months until the gantries could be rebuilt with an extended cantilever over the towpath. In recent years the swing bridge here has in turn been replaced with piping below the road surface, rendering navigation impossible at this point. The entire canal was officially closed to all navigation with effect from 1st January 1963, but in recent years long stretches of it have been reopened for use by pleasure craft. Bainsford presents one of the major problems which will need to be solved if the canal is eventually to reopen right through from the Forth to the Clyde.

Turn left along the landscaped towpath on the far side of the Bainsford crossing and proceed along the north bank for some distance, past the remains of Locks Nos. 6, 7, 8, 9 10 and 11. Beside Lock No. 9 is a swing bridge carrying the railway line to Grahamston Station over the canal. The original swing bridge here was installed when the Stirlingshire Midland Junction Railway, from Larbert to Polmont, was opened in 1850. The present bridge dates from 1899; it no longer swings. Note also, at this point, one of the surviving canal lock-keepers' cottages. A short distance along, on the south bank near Lock No. 10, was Tophill Works, a maintenance depot of the Forth & Clyde Navigation Company. It was here that the famous *Charlotte Dundas*, the world's first practical steam vessel, was moored for a period before being broken up in 1860. At Lock 11, on the south bank, is Rosebank Distillery, which has manufactured whisky here since the early years of the nineteenth century – a surviving canal-side industry no longer canal-served. We have now arrived at Camelon Bridge, and ahead is the distinctive curved gable of 'The Rosebank,' a restaurant opened in 1988, in a former warehouse of Rosebank Distillery.

Cross to the south bank at Camelon Bridge and walk along Glenfuir Road, past Locks Nos. 12, 13, 14 and 15, until eventually arriving at the famous Lock 16, where a short connecting road crosses the canal. It will have been noted by now that most of the surviving lock gates in the Falkirk area are in very poor condition, but at Lock 16, at least, most of the sluice mechanism (known in England as 'paddle' mechanism), for opening the apertures to control lock-chamber water-level, remains in position. Lock 16 was famous in canal terms because it marked the point of junction with the Edinburgh & Glasgow Union Canal, (usually referred to simply as the 'Union Canal') which was opened in 1822 to form a link between the Forth & Clyde Canal and the City of Edinburgh. The point of junction is marked by an indentation in the south bank of the Forth & Clyde, just beyond Lock 16. A nearby car park and stretch of grass marks the site of 'Port Downie', a large canal basin for interchange of traffic between the two canals, which has been filled in. Alongside is the 'Union Inn', built to provide refreshment for Glasgow-Edinburgh canal passengers, at the time of opening of the Union canal, and still functioning as a public house at the present time – a remarkable survival of the canal age. The pub was known locally for many years as 'Auntie Kate's' – in commemoration of this the 'Auntie Kate' name has been given to a little boat which can sometimes be seen moored nearby.

Return to Glenfuir Road and continue in the same direction as before. The road here has been built on the line of the lower part of a sequence of locks which was necessary to overcome a basic difference in level between the Forth & Clyde and Union canals. The locks were filled in and levelled after commercial traffic on the Union Canal ceased in 1933. Turn right into Tamfourhill Road and almost immediately left, through a little wicket-gate, into a grove of trees. It will be immediately apparent that this is no ordinary grove of trees. Ahead is what appears to be a deep ditch of even V-profile, with, however, no water in the bottom. It is in fact a most excellent stretch of the defensive ditch dug by the Romans, as an extra fortification, just beyond the line of the Antonine Wall (which cannot itself be seen here). It survives at this location to a width of about 40ft. and a maximum depth of 15ft. – no doubt close to its original dimensions. This spectacular feature terminates at a property known as 'Watling Lodge', which in

The Union Canal west of Falkirk High Station, showing close proximity of the E & G railway. ***Strathkelvin District Libraries***

turn marks the location of the junction where the Roman military road, which followed an east-west course behind the Antonine Wall, branched off northwards to serve the forts at Camelon, Ardoch, Strageath and Perth.

Return again to Glenfuir Road and continue in a southerly direction. Over to the right notice Greenbank Place, which marks the former course of the road, before the canal locks were filled in and levelled. Ahead is an embankment carrying the Edinburgh & Glasgow Railway, with an underbridge for the line of Glenfuir Road. This bridge was brought into use in 1961 to replace the former Summerford Viaduct, of six arches, demolished at that time. The bridge aperture is on the lateral line (in fact a little to the south) of the principal arch of the viaduct, which accommodated the canal lock-flight, and also a branch railway line. The road, on its former alignment, passed through another arch to the west. To regain this alignment the present road bends sharply to the right on the other side of the bridge, and to the left a short distance further on. Almost immediately the terminal point of the Union Canal comes into view on the left-hand side of the road. A trip boat for disabled passengers, the *Govan Seagull* belonging to the Seagull Trust, is sometimes moored at this location. The top of the lock-flight from the Forth & Clyde Canal was in fact some 570 yards east of here, but the high-level extension from the top-lock westwards (known as Port Maxwell) was opened in 1823 to reduce walking time between the two canals – passengers were required to make the link on foot. Turn along the towpath on the north bank of the Port Maxwell cut and watch for signs of the lock-flight alignment coming into view on the left. A short distance from the junction between the top of the lock-flight and the 1823 cut, through a gap in the trees, the stone-built remains of one of the locks will be observed, adjacent to a picnic site. Continue eastwards along the canal bank, noting the E & G railway converging from the west, a short distance along, to run alongside the canal for some way. This is an excellent vantage point for railway photography, although the variety of locomotive and train types to be seen here is extremely limited nowadays. Continue through a typical Union Canal overbridge until the west portal of Falkirk Canal Tunnel comes into view. Leave the canal-side by a footpath which diverges a short distance before the tunnel. A railway underbridge leads to Drossie Road and FALKIRK HIGH STATION car-park, immediately to the right. (WALKING TIME, FALKIRK HIGH STATION TO FALKIRK HIGH STATION, VIA FALKIRK TOWN CENTRE, BAINSFORD, CAMELON BRIDGE AND SUMMERFORD: APPROXIMATELY 3 HOURS).

Alternative Walk. Leave FALKIRK HIGH STATION by the exit near the eastern end of the up platform and turn right across the overbridge. Continue along Slamannan Road, noting the entrance to Falkirk Canal Tunnel below the road on the right-hand side. A short distance further on cross to Princes Park on the other side of the road. At a high point of this park, on a solidly-built pedestal, will be found a diorama of the very wide range of hills and mountains which can be seen on a clear day, looking northwards from this location. The diorama was installed as long ago as September 1935, with funding from the *Daily Record* newspaper. The town of Falkirk can also be viewed to good advantage from Princes Park, which is 375ft. above sea level. Leave the park and continue along Slamannan Road, turning right into Lochgreen Road a short distance further on. Continue until another public park is seen alongside the right-hand side of the road, then turn right at the next junction into Greenbank Road. Almost immediately will be seen the impressive monument to the second Battle of Falkirk, which was fought out close to this point. On 17th January 1746 the Jacobites, under Lord George Murray, inflicted a resounding defeat on the Hanoverian army. Only three months later, as is well known, this result was reversed at Culloden and Bonnie Prince Charlie's pretension to the British throne was at an end. The Second Battle of Falkirk was the penultimate battle of the 'Forty-Five' rebellion, and as such it was the second-last land-battle fought on British soil. After inspecting the inscription on the monument continue along Greenbank Road, and after a few twists and turns the terminus of the Union Canal at 'Port Maxwell' will be reached. Return along the north bank of the canal to FALKIRK HIGH STATION, as described in the section relating to the principal Falkirk walk, above. (WALKING TIME, FALKIRK HIGH STATION TO FALKIRK HIGH STATION, VIA 'BATTLE OF FALKIRK' MONUMENT: APPROXIMATELY 1 HOUR).

Linlithgow Town Trail

Linlithgow is one of the key historical locations in Scotland. Its famous palace was the birthplace of Mary, Queen of Scots and is of prime significance, but there is much of interest in the town besides this important structure. The Palace is perhaps not seen to best effect form the train – it is masked to some extent by the notable St. Michael's Parish Church, with its startling spire of laminated timber and aluminium, erected in 1964 at the instigation of The Very Rev. Dr. David Steel (father of the famous politician of the same name) to replace an open stone crown which had been removed for reasons of safety in 1821. During the mediaeval period Linlithgow was one of the most important towns in Scotland. In 1368 it was established as a member of the Court of the Four Burghs – the 'Convention of Scottish Local Authorities' of the period –

joining Edinburgh, Stirling and Lanark. Together these four towns assumed custodianship of the Scottish standard weights and measures – Linlithgow, for its part, held the definitive 'firlot', by which grain was measured.

Much of the character of Linlithgow as an ancient burgh has been preserved (despite some alarming redevelopment of the 1960s, which billboarded a few dramatic question marks). The long, winding character of the old High Street has been maintained, and some interesting town houses of the post-mediaeval period survive. A notable feature is the number of old 'well wynds' leading down to the High Street from the southern hillside, each one to a well of the appropriate name. St. Michael's Well survives as a good example. Paradoxically, the railway, when it arrived in 1842, had a *beneficial* effect, with regard to preservation of the old town. This was because it was built along the line of the southern boundary of the burgh, previously marked by a continuous line of back property walls, or 'yardheids'. It accordingly restricted lateral development, so reinforcing the linear nature of the town and perpetuating the fundamental importance of the High Street as a thoroughfare.

Leave LINLITHGOW STATION by the main entrance on the north side, and immediately turn left through a tunnel under the railway. Follow the road round to the right, up a short hill, and soon a dome-like stone structure will come into view, at the rear of a small public garden. This is a *doocot* or dovecote, of a type common in different parts of Scotland. It is of considerable antiquity, dating perhaps from the sixteenth century. Its purpose was to provide a supply of fresh pigeon meat during the winter months, at a period when salted beef from cattle traditionally slaughtered in the autumn became progressively more unpalatable as the season went on. This one has about 400 nest holes inside and three projecting stone string-courses around the outside, supposedly to stop rats from climbing in.

Straight across the road from the doocot is Manse Basin on the Union Canal, frequently bustling with activity during the summer months. Serious traffic on the Canal ceased during the 1930s, but ironically it was not officially closed until 1965, and not long afterwards the first moves were made towards reopening it for recreational purposes. During the 1970s various pleasure boats were introduced onto the canal. These included, in 1978, the attractively-styled little motor vessel *Victoria*, which can carry up to twenty passengers (twelve in acceptable comfort!) on short excursions from Linlithgow. Other vessels now regularly to be seen at Manse Basin include the *Janet Telford*, another trip boat operated by the Linlithgow Union Canal Society. Also at the Basin, in former Canal stables, is the Linlithgow Union Canal Society's museum, open during the summer months only. Nearby canal features include a typical Union Canal bridge, with deep grooves cut in an inner corner of the north abutment by generations of contact with barge tow-ropes, and Canal House, on the south bank west of the museum where engineer Hugh Baird is said to have lived during construction of the Canal. Also of great interest is a row of Edinburgh & Glasgow Railway stone sleepers, forming the edge of the north quay-side. There were regarded as very old-fashioned technology when the railway opened in 1842 and were soon replaced with the conventional wooden variety. They survive in the structure of walls, loading banks, etc., along the entire length of the railway. Their presence here reflects the fact that the Union Canal was purchased by the Edinburgh & Glasgow Railway Compny in 1849. Each sleeper has a rectangular depression, countersunk to accommodate an iron chair. Within the rectangles are holes for wooden plugs, into which securing spikes were driven.

Monument commemorating the Battle of Falkirk, 1746. ***Strathkelvin District Libraries***

Continue westwards along the road on the north bank of the canal (known here as Strawberry Bank) until the railway comes alongside once again. On the left is an interesting row of houses, known as Royal Terrace, associated with the coming of the railway. A short distance along is a neat little cast-iron footbridge across the line – its shallow arch seems original, perhaps of 1840s date. Cross the footbridge and notice immediately the impressive retaining wall on the north side, at the head of Lion Well Wynd. Environmental improvement here won a Saltire Society award, in 1979. Descend the picturesque Wynd, turn left at the foot, into West High Street, and continue for a distance until an ancient three-storey crowstepped building comes into view, high above the road on the left-hand side. This is West Port House and is of special interest because it marks the site of the 'west port' or west gate of the burgh during mediaeval times. It was built as a 'country house' for the local Hamilton family in the sixteenth century, and survived as a family residence until 1988, when a former provost of Linlithgow moved out. It was subsequently converted to flats.

Return to the foot of Lion Well Wynd, and continue eastwards along West High Street to The Cross of Linlithgow. Of immediate note is the Cross Well, in the centre of a little square. The Well was comprehensively rebuilt in 1807, the work of an Edinburgh stone mason, Robert Gray, who worked with a mallet strapped to his right

West Port House, Linlithgow. ***Strathkelvin District Libraries***

arm, having no right hand. Its elaborate decoration includes a unicorn supporting the National Arms of Scotland at the very top, but perhaps more importantly a representation of a town drummer, a familiar figure in Scottish burgh life. At the rear left-hand corner of the square is the 'Cross House' an impressive three-storey town house of traditional style, built about 1700 by Andrew Craufurd, a prominent local burgess. To the right of it is the 'Town House' which has a complex history. It has been the subject of periodic remodelling (the double flight of steps in front, for example, dates only to 1907), although the main substance of the three-storey building dates to 1668-70. It replaced an earlier Tolbooth destroyed by Oliver Cromwell. In front of the Tolbooth, it is recorded, stood a market cross with two small shops at the base. Within the 'Town House' is located the Linlithgow Tourist Office, where an excellent range of publications is available to facilitate exploration and understanding of the district.

Step across the road from the Cross Well and notice that at No.79 High Street a fire mark of the Sun Insurance Company can be seen high up on the building frontage. In the event of a fire, this enabled the company's firemen to identify the building as 'one of their own', although there is evidence that firemen of the different insurance companies were eventually allowed to co-operate with each other. Once common, these marks are now very scarce on buildings in Scotland. Continuing eastwards along High Street, notice the picturesque property at Nos. 40-48, known as the 'Hamilton Lands'. The gables are turned to the street in a style expected in Holland, but which was once also very common in Scottish burghs. This four hundred-year-old tenement was restored by the National Trust for Scotland in 1958. Just across the road from the Hamilton Lands, in front of St. Michael's Hotel is a figure of St. Michael, mounted on a solid stone base with the inscription 'SAINT MICHAEL IS KINDE TO STRAINGERS'. This represents the remains of St. Michael's Well, one of Linlithgow's famous series of High Street wells, this one constructed by the town in 1720. A few more steps in an eastward direction takes us back to LINLITHGOW STATION, so completing our circular tour of the old burgh. (WALKING TIME, LINLITHGOW STATION TO LINLITHGOW STATION, VIA CANAL BASIN, LION WELL WYND, WEST PORT HOUSE AND THE CROSS WELL: APPROXIMATELY 1 HOUR).

When in Linlithgow, time should also be allocated to visit LINLITHGOW PALACE and ST. MICHAEL'S PARISH CHURCH, both of which are reached by way of Kirk Gate, which begins at the Cross Well. Detailed guidebooks to both are available. Even if time is not taken to visit the interior of the Palace much of interest can be seen by taking a walk round the outside. At the top of Kirk Gate is the massive gateway known as the 'Outer Entry', dating from about 1535, with its colourful representation of the arms of the four orders of chivalry to which James V, the monarch responsible for the gateway's construction, belonged. James V was also responsible for the creation of the main entrance to the palace building (on the south side) still in use today. Remains of an earlier entrance, dating from construction of the castle in the reign of James I of Scotland (c. 1424-37), can be seen on the eastern side of the building, surmounted by a representation of the royal arms of Scotland. Also on the eastern side, at a lower level, is the remains of a barbican or 'outer bulwark', dating from the reign of James IV (1488-1513); it consists of two round towers, with a smaller stair tower between, connected to the main building by flying buttresses. Nearby, it is worth noting a flat stretch of ground beside the loch, known as the 'Bow Butts', where archery was practised. Inside the Palace there is much of interest to see, but note especially the courtyard fountain (installed about 1538); the Great Hall in the east range (at First Floor level), with its massive stone fireplace at one end; and the King's suite of apartments in the west range (also at First Floor level), which includes an outer hall, a 'presence chamber' or throne room, a bed chamber, and a small private closet, or relaxing room, with a large bay window to give a good view across the loch. The bay window can also be seen to good effect on the outside of the building (north side). St. Michael's Church, adjacent to the Palace, is claimed to be one of the finest parish churches in Britain. Its outstanding feature is window tracery, especially that on the notable St. Katherine's Aisle window, south transept.

Station Linking by the Union Canal

On 23rd July 1990 an important initiative relating to the environment of the Edinburgh & Glasgow line was launched at a ceremonial tree-planting by Ian Lang, then Minister of State at the Scottish Office, at Polmont Station. This was the Green Link Project, its purpose being to improve the appearance of stations and adjacent railway land along the route, and responsibility for its implementation was vested in the Central Scotland Countryside Trust. As part of the Project prominent sign boards have now been erected at Falkirk High, Polmont and Linlithgow stations, giving clear instructions for recreational walking between them, along the towpath of the Union Canal. The Falkirk High board, for example, has the following wording: 'GREEN LINK CANAL WALK. TO POLMONT STATION via tunnel (1½ hours – TAKE A TORCH) and LINLITHGOW STATION via tunnel and aqueduct (4 hours). Cross car park, down steps, under subway. Turn left onto towpath. RETURN BY TRAIN'.

These walks can be undertaken exactly according to the directions on the notice boards. They can be accomplished in either direction, with the options of a 1½-hour walk between Falkirk High and Polmont, a 2½-hour walk between Polmont and Linlithgow, or a 4-hour walk for the full distance between Falkirk High and Linlithgow. To gain the towpath from Polmont Station leave the station on the up side and turn left along a minor road for some distance, then turn left again, across the railway by a substantial stone bridge, now used only as a footbridge. At Linlithgow Station the towpath is reached by leaving the main station entrance, turning immediately left under the railway, and following the road round to the right until Manse Basin comes into view (exactly as in the 'Linlithgow Town Trail').

There is much of interest to see. Between Falkirk and Polmont a most unusual experience is to walk through the 700 yard Falkirk Canal Tunnel. The towpath is in good repair throughout, and a stout handrail separates the pedestrian-way from the water course, but the tunnel is dark and damp, so torches and waterproofs are essential items of equipment. Construction of a tunnel through Prospect Hill was necessary because of the local landowner, William Forbes, did not want the canal traffic to disturb the view from his mansion at Callendar House – he went so far as to commission an artist's impression of what the effect would be. However, it created an exciting experience for the early canal voyagers, as witnessed by the compiler of *A Companion for Canal Passengers betwixt Edinburgh and Glasgow* in 1823:

> The passengers on their way from Edinburgh, will have seen with surprise, the aqueducts, and other various works, which have been constructed on the line of the Canal; but when *here*, they see the wide chasm, and the distant light, glimmering

through the lonely dark arch of nearly half a mile in length, they are struck with feelings of awe; and, as they proceed through it, and see the damp roof above their heads, – feel the chill rarified air, – and hear every sound re-echoing through the gloomy cavern, – their feelings are wound to the highest pitch. At the same time, there is something uncommon in the idea, that *there,* secluded from the light of day, they are silently gliding through the bosom of a hill, whose surface is cheered by the enlivening rays of the sun, and covered with crops, cattle, and the cheerful dwellings of man.

To understand the full impact of this experience at Falkirk, it is necessary to bear in mind that at the time there were no other canal or railway tunnels in existence anywhere in Scotland. The tunnel water course was reopened to passengers in July 1988, when the Seagull Trust's *Govan Seagull* was the first boat through, so it is now possible to re-live the experiences of 1823, afloat as well as on foot.

A short distance beyond the eastern portal of the tunnel is Glen Bridge, one of the highest overbridges on the Union Canal. Carved on the two key-stones of the arch, one on either side, are two grotesque faces, one apparently smiling and the other scowling. Accordingly, the bridge is sometimes referred to as the 'Laughing and Greeting Bridge'. The scowling face looks westwards, and is supposed to depict the contractor contemplating the difficult task of tunnel construction ahead. The smiling face looks eastwards to a comparatively easy, level stretch of the canal. The stonework carving also includes the date 1821 and the bridge number 61, a consistent feature on Union Canal bridges. Bridges from here to Linlithgow are numbered from 61 to 43, in descending order.

Between Polmont and Linlithgow the most signifcant features are the Slamannan Basin, at Causewayend, the Avon Aqueduct, and Woodcockdale Stables. The Slamannan Basin, about two and a half miles east of Polmont, on the south side of the canal, was

Union Canal Basin at Linlithgow. ***Strathkelvin District Libraries***

the terminal point of the 4ft. 6in. gauge Slamannan Railway, opened in 1840. At this date an 'inter-city' passenger service was commenced between Glasgow and Edinburgh, by way of the Garnkirk & Glasgow, Monkland & Kirkintilloch, Ballochney and Slamannan railways (all of which were of the same 'narrow' gauge), thence from Causewayend Basin to Edinburgh by means of a track-boat on the Union Canal. This remarkable means of crossing between the cities was only available for a very short time, being totally superseded by the Edinburgh & Glasgow Railway Company's through line in 1842. Even before then, however, the Causewayend-Edinburgh canal leg and succumbed to stagecoach competition between these places. Now silted up and overgrown, the 150ft. square basin still survives as a memorial to an interesting episode in Edinburgh-Glasgow transport history.

About half-a-mile east of the Slamannan Basin is the remarkable Avon Aqueduct, nearly 300yds. long and towering 85ft. over the River Avon below. It was designed by the Union Canal engineer Hugh Baird, in consultation with the famous Thomas Telford, and is the largest of three major aqueducts on the Union Canal. It has twelve stone arches, and the water is carried across the top in an iron trough. It has an outlet for spectacular discharge of excess water into the river below – a useful facility, also, if the aqueduct requires to be emptied of water for purposes of maintenance. A short distance to the east of the aqueduct, close to a road bridge, are Woodcockdale stables, once occupied by canal tow-horses though now used by local scouts. Between Edinburgh and Falkirk the canal was divided into four stages of about eight miles each, for purposes of changing horses and determining tolls. Woodcockdale marked the end of the third stage from Edinburgh and the beginning of the fourth (and last), the earlier stage boundaries, to the east, being at Ratho and Winchburgh. Curiously, however, the stone pillar, serving as a marker between the third and fourth stages, is at the west end of the Avon Aqueduct. Nearby is a stone milepost – one of the regular features on the Union Canal for which a vigilant eye should always be looking. Others include numbers incised on the bridges and grooves in stonework caused by the rubbing of tow-ropes.

Backward Glances

GLASGOW QUEEN STREET STATION. The arches on the left of the above view (of 1943) formed the original 1842 entrance to Queen Street station. They were demolished during modernisation in the late 1960s. The building on the right, originally the 'Wardlaw Kirk', served as the Edinburgh & Glasgow Railway main offices for a period. The view below shows LNER 'Pacific' locomotive No 2566 *Ladas* departing with an Edinburgh train in April 1938.
A.A. Maclean Collection / C. Lawson Kerr Collection, Mitchell Library

COWLAIRS INCLINE. **The above view shows Class A2 4-6-2 locomotive No. 60535 *Hornets Beauty* climbing Cowlairs Incline on the 'Queen of Scots' Pullman train in July 1957 banked by an 0-6-2 tank locomotive. The lower view shows the bankhead during the cable-hauled era, c.1893, with the steam winding-engine house on the left.** ***C. Lawson Kerr Collection, Mitchell Library / Mitchell Library***

EASTFIELD SHEDS. Above is a general view of the early 1950s, showing a great variety of locomotive types 'on shed'. The lower picture, dating from the same era, shows Class D30 4-4-0 No. 62418 *The Pirate*, formerly a well-known performer on Glasgow-Edinburgh trains, although by this date reallocated to Thornton Junction and working between Fife and Glasgow. *C.J.B. Sanderson / C. Lawson Kerr Collection, Mitchell Library*

BISHOPBRIGGS STATION. The upper view shows the station as it was in May 1961, before the extensive reconstruction of later years. The lower picture is of a Kirkintilloch-bound local train, hauled by Class V1 2-6-2 tank locomotive No. 67664, on 24th April 1961.
D Martin / W.S. Sellar

LENZIE JUNCTION. The upper picture shows *The Pirate* again, this time about twenty years earlier, on a Glasgow-Edinburgh train. The lower view is of the 'up' platform at Lenzie, during the Edwardian era. The upper part of the signal box was later extended out over the platform. *Tom Lindsay / Strathkelvin District Libraries Collection*

DULLATUR & CASTLECARY. Both of these photographs were taken on the same day, 20th May 1961. The upper view is of Dullatur Station, looking east, with a 'Black 5' 4-6-0 locomotive passing through on a goods train. The lower view is of Castlecary Station, also looking east. Both stations were closed to passengers in 1967. ***D. Martin (both)***

SUMMERFORD VIADUCT, CAMELON. **Summerford (or Camelon) Viaduct was demolished in 1961 and replaced by a new embankment and bridge, immediately to the south. The above picture shows the viaduct from the south-east, in April 1955, with the Camelon goods branch passing under the main arch. The view below shows the main arch shortly before demolition, with the new bridge visible beyond.** ***The late J.F. McEwan / D. Martin***

FALKIRK HIGH STATION (I). The old buildings at Falkirk High were replaced by new ones, of modern construction, in 1980. Above is a general view of the station, looking east, in March 1958. Below is a Glasgow-bound local train approaching the 'down' platform on 11th May 1961. ***The late J.F. McEwan / D. Martin***

FALKIRK HIGH STATION (II). The upper view shows Class D49 4-4-0 locomotive No. 62704 *Stirlingshire* approaching Falkirk High on a Glasgow Queen Street - Kirkcaldy train on 18th February 1956. The lower picture shows Class D11 4-4-0 No. 62671 *Bailie Macwheeble* leaving Falkirk High with a Glasgow local train on the same day. *R.C. Nelson (both)*

POLMONT ENGINE SHEDS. Polmont Shed was once a familiar sight on the down side of the line, about half way between Edinburgh and Glasgow. Above is a general view of the east end of the shed, on 11th May 1961. Below is No. 62693 *Roderick Dhu* and two other members of the 'Director' 4-4-0 Class, long familiar on Edinburgh-Glasgow trains, in store on a siding at the east end of Polmont Shed yard, on 18th April 1960. *D. Martin (both)*

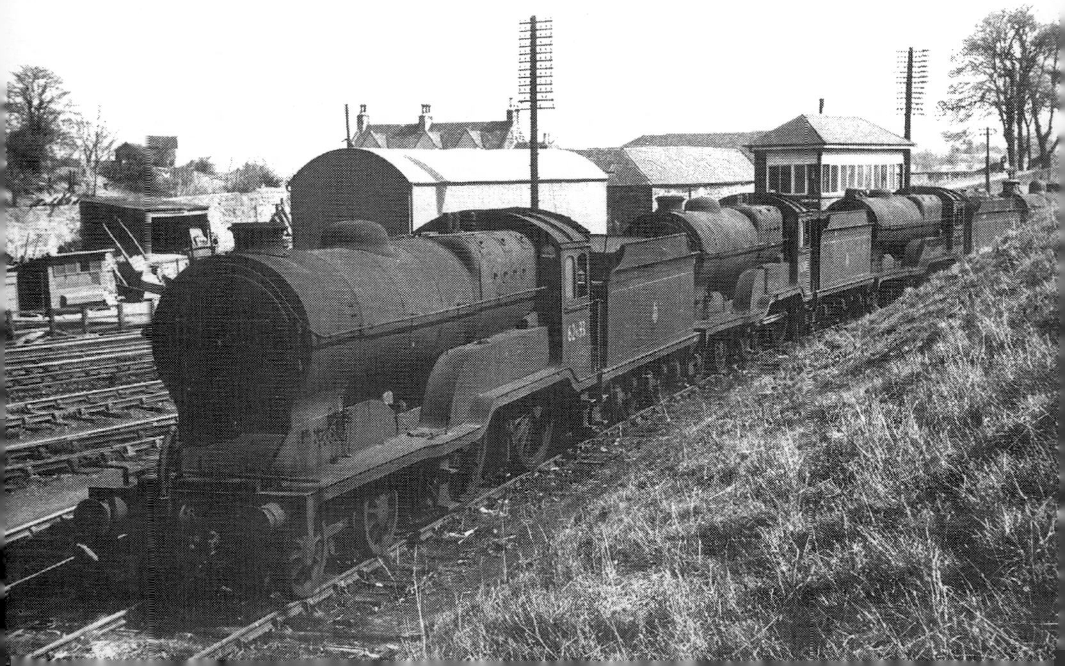

POLMONT STATION. The top photograph shows Class A4 streamlined 4-6-2 locomotive No. 60024 *Kingfisher* at Polmont on a Glasgow-London King's Cross train on 11th June 1960. The lower photo shows Class C16 4-4-2 tank locomotive No. 67488 in the Bo'ness Branch bay platform at the east end of the station, on 9th June 1957. *D. Martin / W.S. Sellar*

MANUEL STATION. Until 1930 Manuel Station was served by 'Low Level' trains on the former Monkland Railways system, as well as by Edinburgh-Glasgow line trains at 'High Level'. The 'Low Level' platform was at right angles to the 'High Level'. After the Monklands-Bo'ness trains ceased in 1930 the booking office on the lower level (below) remained in use for 'High Level' trains until 1967, when they too succumbed. The above view is of the waiting room on the up 'High Level' platform. *A.A. Maclean (both)*

LINLITHGOW STATION. **The above view of Linlithgow Station is by the famous early photographers D.O. Hill and Robert Adamson. Dated about 1844, it is possibly the world's earliest railway photograph. The lower photo is of Class A4 4-6-2 No. 60004** ***William Whitelaw*** **at Linlithgow with the 4.00 pm Glasgow Queen Street - Leeds, on 30th August 1956.** ***Scottish National Portrait Gallery (Scottish Photography Archive) / R.C. Nelson***

HAYMARKET STATION (I). These views both show Haymarket Station as it was before the extensive rebuilding of the early 1980s. Above is the main island platform. Below is the interior corridor and stairs which led to the platforms. *ScotRail (both)*

HAYMARKET STATION (II). These views show the original part of the station, dating from 1842. The upper photograph shows the eastern end of the original train shed, where it joined the main stone-built office block. The train shed was removed to Bo'ness, for preservation, in 1982. The lower view is of the main office block, at street level, during the 1950s. *A.A. Maclean / The late J.F. McEwan*

PRINCES STREET GARDENS. The above view shows North British Railway '317 Class' 4-4-0 locomotive No. 320 on a down express, with 'The Mound' and one of the Gardens footbridges in the background. The lower photo shows LNER Class D29 4-4-0 No. 9359 *Dirk Hatteraick* on an up semi-fast in 1938. *Photomatic (both)*

EDINBURGH WAVERLEY STATION (I). The above view shows a 'modernisation plan' Swindon-built diesel set approaching the station from Glasgow on 1st June 1959. The lower view is of a scene recently familiar, but now superseded by the ubiquitous ranks of Class 158. Class 47/7 diesel locomotive No. 47712 *Lady Diana Spencer* stands at Platform 14 with a train from Glasgow on 7th September 1983. *ScotRail / D. Martin*

EDINBURGH WAVERLEY STATION (II). **John Menzies bookstalls are a feature of the railway station scene all over the U.K. They have had a presence at Edinburgh Waverley since 1862. Over the years the style of their bookstalls has changed many times. These two Waverley examples are from the 1930s (top) and the 1960s (bottom).** ***John Menzies (UK) Ltd. (both)***

FALKIRK GRAHAMSTON STATION (I). The upper photograph is of Class C15 4-4-2 tank engine No. 67472 on a local service, 21st January 1956. The lower view is of Class V3 2-6-2 tank locomotive No. 67669 leaving Grahamston on an Edinburgh Waverley-Larbert service, 5th October 1956. ***R.C. Nelson (both)***

FALKIRK GRAHAMSTON STATION (II). The old station buildings at Grahamston were taken down during 1985-86 and replaced with new ones of ultra-modern design. These two photos are of the down platform building. The exterior view (above) is dated 21st May 1983, and the platform shot (below) a month later, 22nd June 1983. Note the independently-run station bookstall. *D. Martin / Strathkelvin District Libraries*

Recent Events

INAUGURATION OF MODERNISED FALKIRK HIGH STATION. The rebuilt station at Falkirk High was officially opened on 8th December 1980 by Councillor James Anderson, Convenor of Central Regional Council and a qualified train driver with BR. He is pictured at the ticket office (above), and preparing to drive a train to Glasgow (below), in each view accompanied by Mr. Leslie Soane, General Manager, BR Scotland. *ScotRail (both)*

INAUGURATION OF MODERNISED FALKIRK GRAHAMSTON STATION. Councillor Anderson again officiated at the opening of the rebuilt Falkirk Grahamston Station, on 28th April 1986. In the above view he is pictured with Provost John Docherty of Falkirk District Council and Mr. Jim Cornell, ScotRail General Manager, at the naming of locomotive 47636 after *Sir John de Graeme*, the 'local hero'. In the lower view he flags a train away. *ScotRail (both)*

INAUGURATION OF MODERNISED LINLITHGOW STATION. The completion of the modernisation of Linlithgow Station, on 4th November 1985, was marked by the naming of locomotive 37401 after *Mary Queen of Scots*, who was born at Linlithgow Palace. Pictured above are Mr. Viv Chadwick, Regional Operations Manager, ScotRail, Colonel William Dalziel, Chairman of the Transport Users Consultative Committee for Scotland, and six year old Leighann Dodds, in appropriate dress. In the lower view a commemorative plaque is unveiled by Tam Dalyell, the well known local M.P. *ScotRail (both)*

REPLACEMENT OF FOOTBRIDGES IN PRINCES STREET GARDENS. One of the old-style footbridges in Princes Street Gardens is pictured during its last days, on 18th August 1990 (above). The new-style wooden bridge, with toughened glass viewing panels for children, is opened on 22nd January 1991 by six year old Flora Munro (below). *D. Martin / ScotRail*

INAUGURATION OF GREEN LINK PROJECT. **A project known as 'Green Link' to improve the appearance of stations and adjacent railway land was launched by Ian Lang, M.P., then Minister of State at the Scottish Office, at a tree-planting ceremony at Polmont Station on 23rd July 1990 (above). One of the 'Canal Walk' notices, erected at stations in connection with the Green Link Project, is pictured at Falkirk High (below).** ***ScotRail / Strathkelvin District Libraries***

Today's Trains

Class 158

Nowadays almost all of the through passenger services between Edinburgh and Glasgow are worked by diesel multiple units of Class 158. Trains are normally two, four or six cars long, made up of two-car sets coupled together.The Class 158s were first introduced onto the route on Monday 17th September 1990, with a train made up of sets 158708, 158709 and 158711 working the 7.00 Edinburgh-Glasgow and remaining in service throughout the day. A full Class 158 timetable was introduced the following month. These were the very first 158s to go into service – since then use of the class has spread all over Great Britain.

The units are fitted with 350 b.h.p. Cummins NTA 855 R1 turbocharged engines, one per vehicle. Transmission is by Voith T211RZ unit, driving two axles via Gmeinder final drives. Maximum speed is 90 mph.

The cars are of aluminium construction, manufactured by BREL Ltd at Derby. They have swing plug, bi-parting power-operated doors. Seating is provided for 66-70 persons in each car, with a mixture of face-to-face and face-to-back seats; one vehicle in each two-car set has a compartment with 15 seats set aside for First Class passengers. The vehicles are air-conditioned..

Classes 150 & 156

Class 150/2

Local trains on the route are worked by multiple units of Class 150/2 and Class 156. Class 150/2 works between Edinburgh and Dunblane and Edinburgh and Bathgate, also on other local services in the Edinburgh area. Class 156 works between Glasgow and Dunblane, and Glasgow and Falkirk Grahamston. The classes were introduced into Scotland in 1987 and 1988 respectively, and have now virtually ousted all the older types of diesel multiple unit. Like Class 158 they form two-car sets, and have Cummins engines (though of 285 b.h.p.), Voith transmission and Gmeinder final drives. Unlike the Class 158s they are restricted to 75 mph.

Class 156

Mile by Mile

1. Glasgow – Lenzie

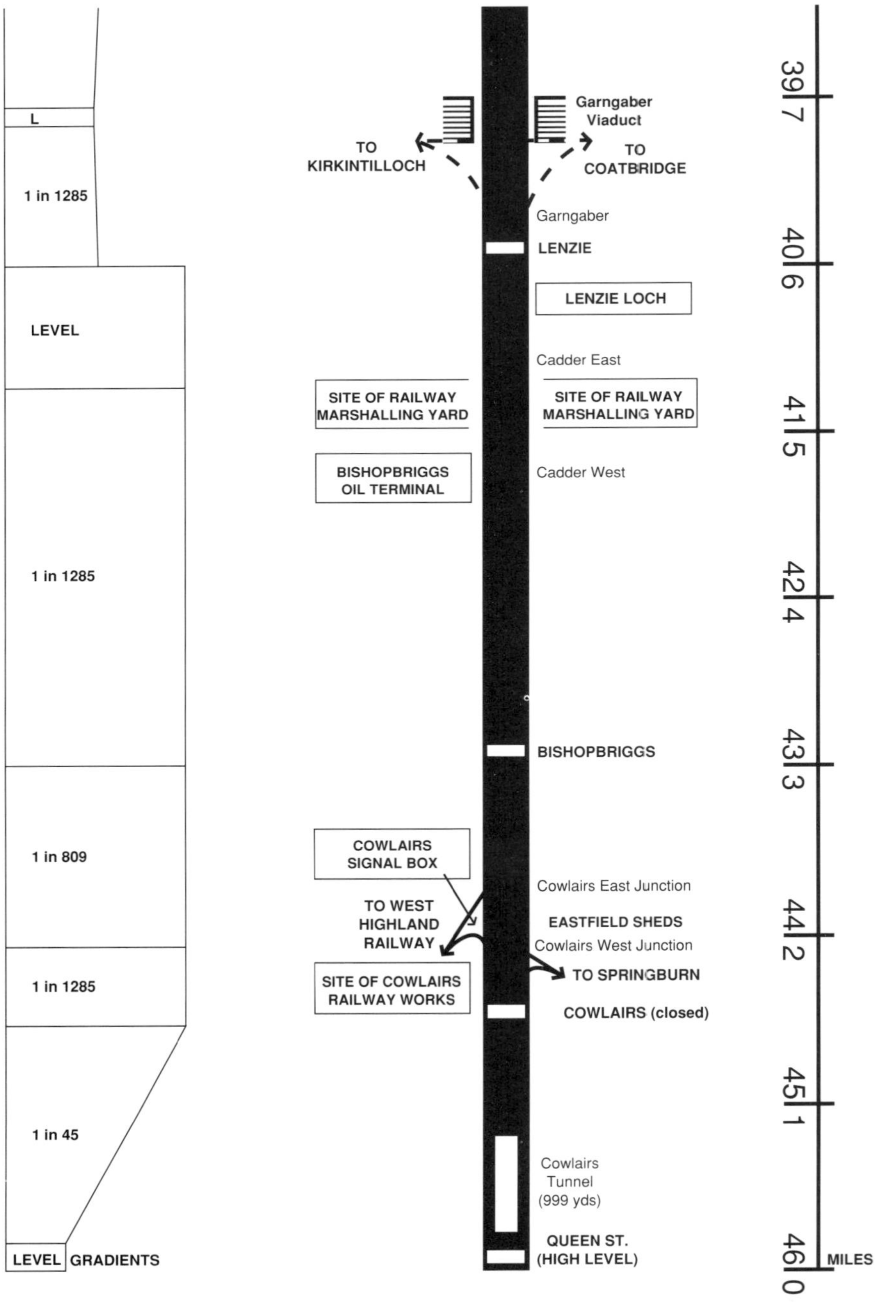

2. Lenzie – Dullatur

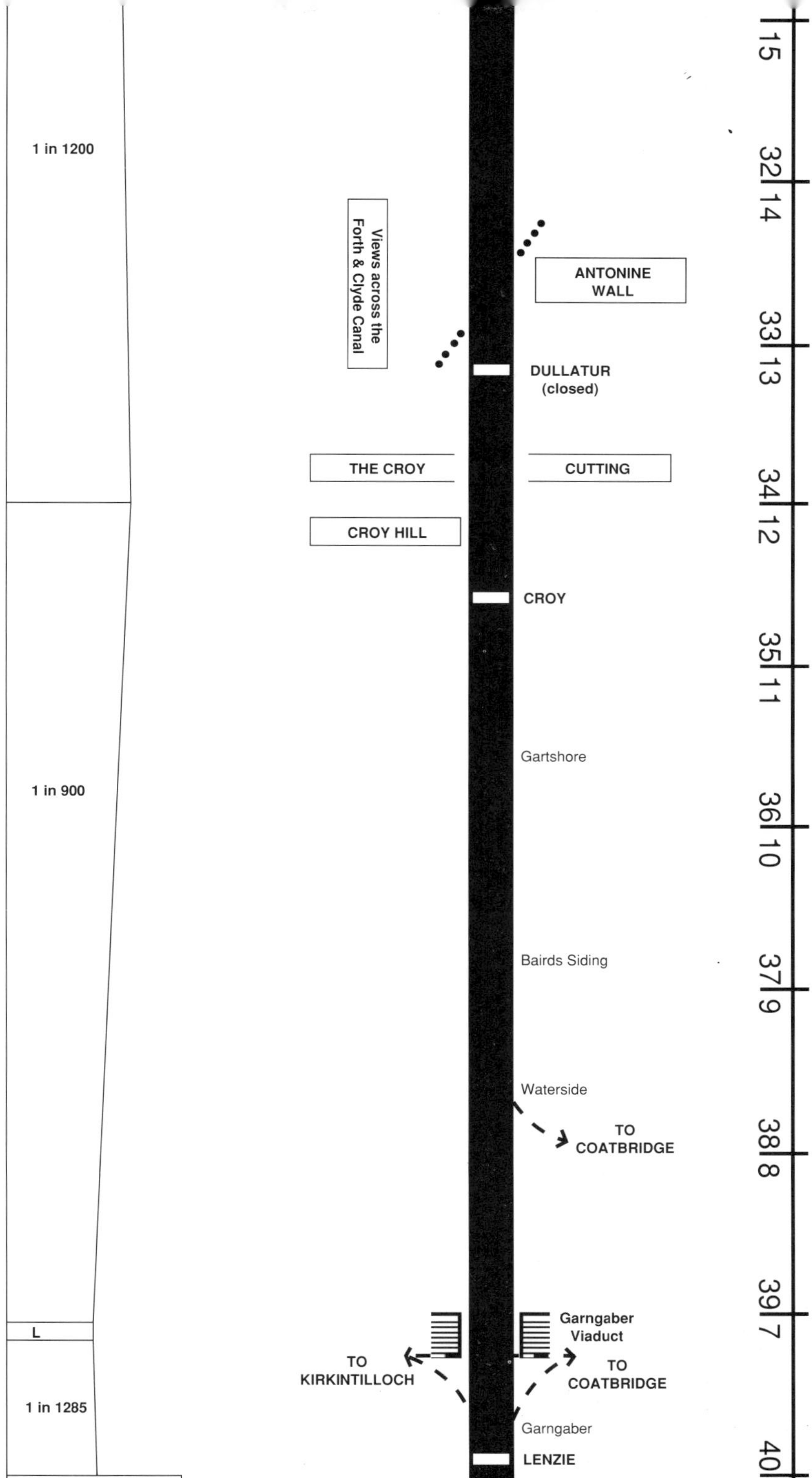

3. Dullatur – Bonnybridge

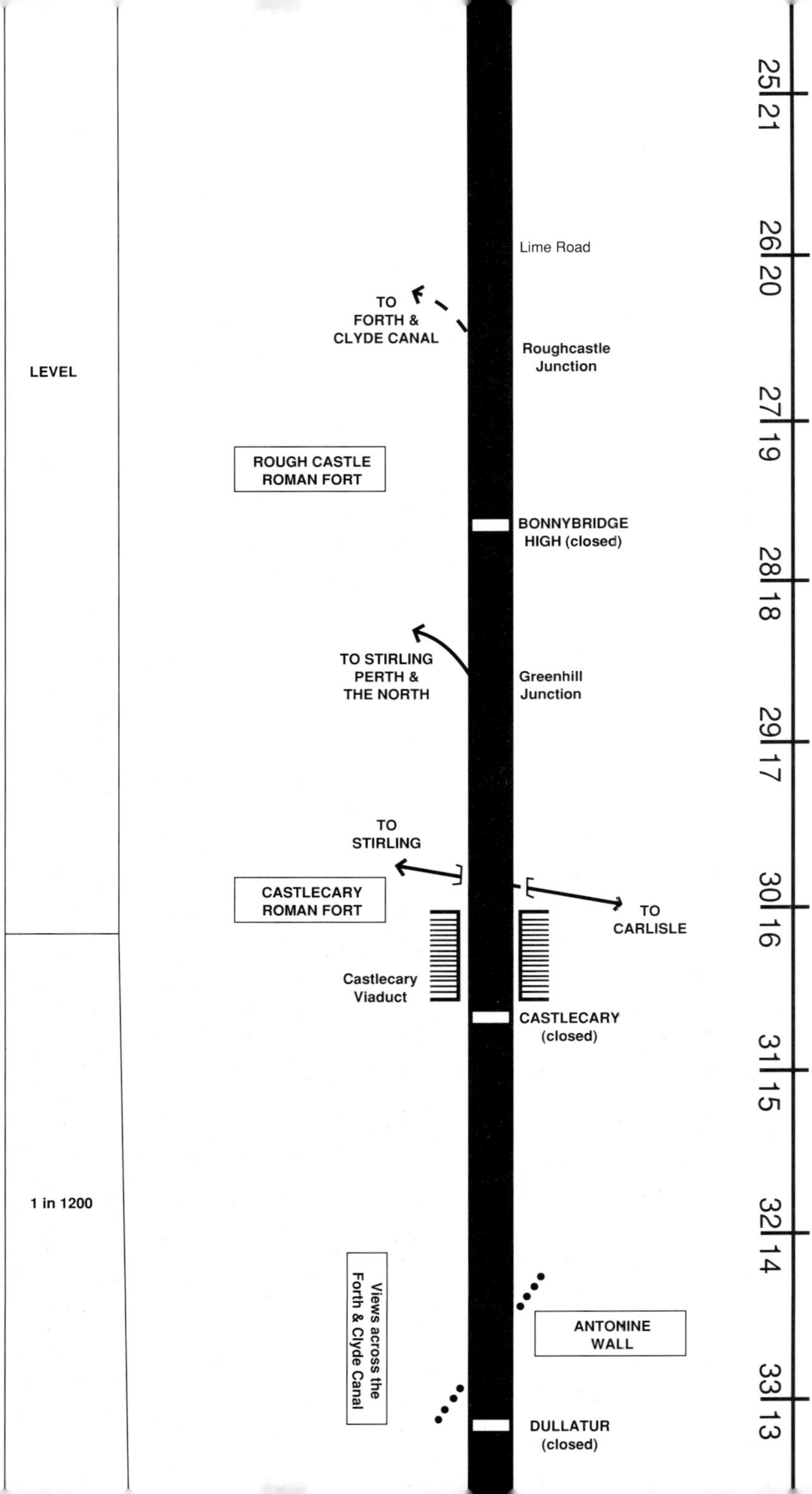

4. Bonnybridge – Polmont

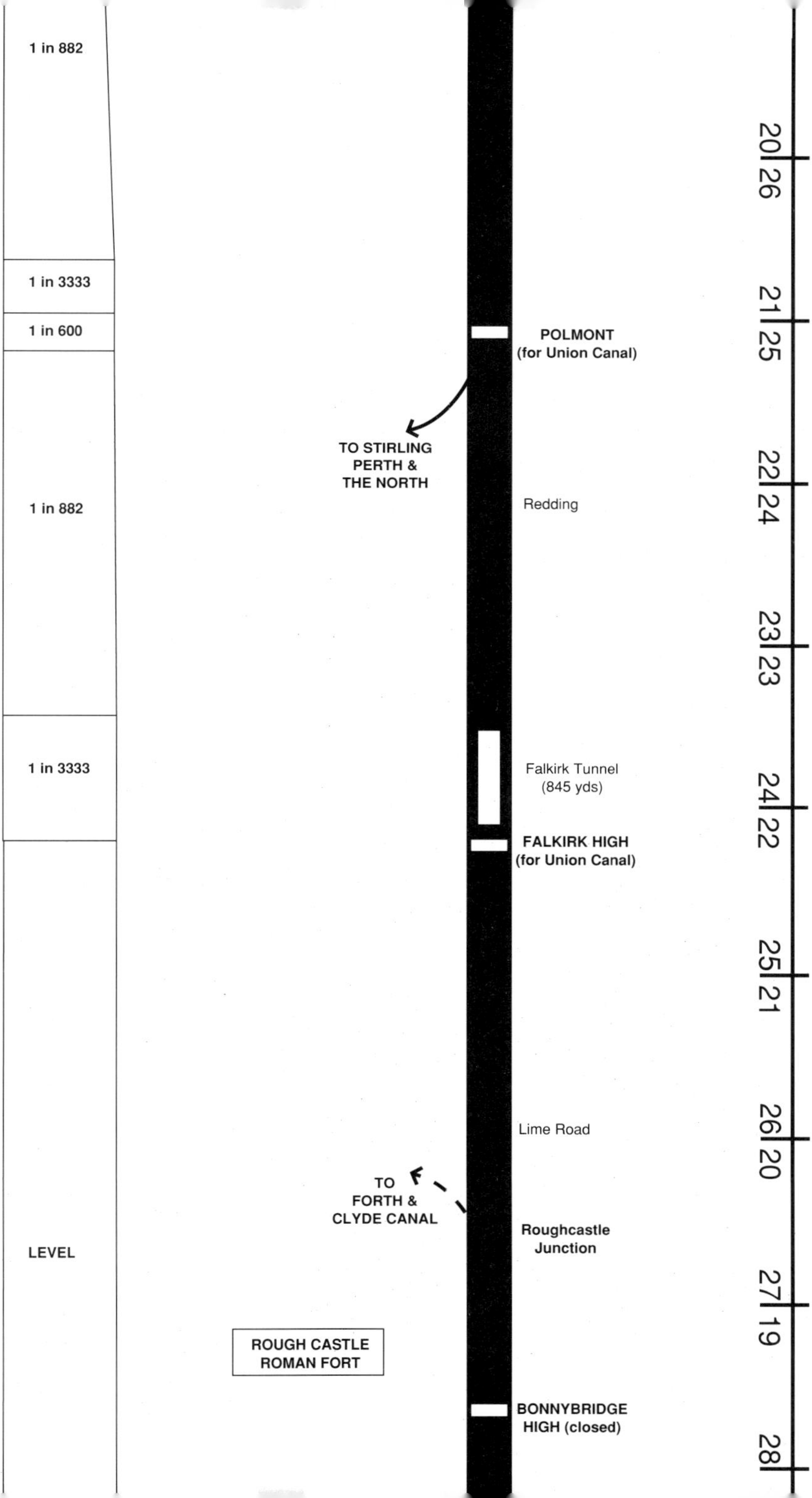

5. Polmont – Linlithgow

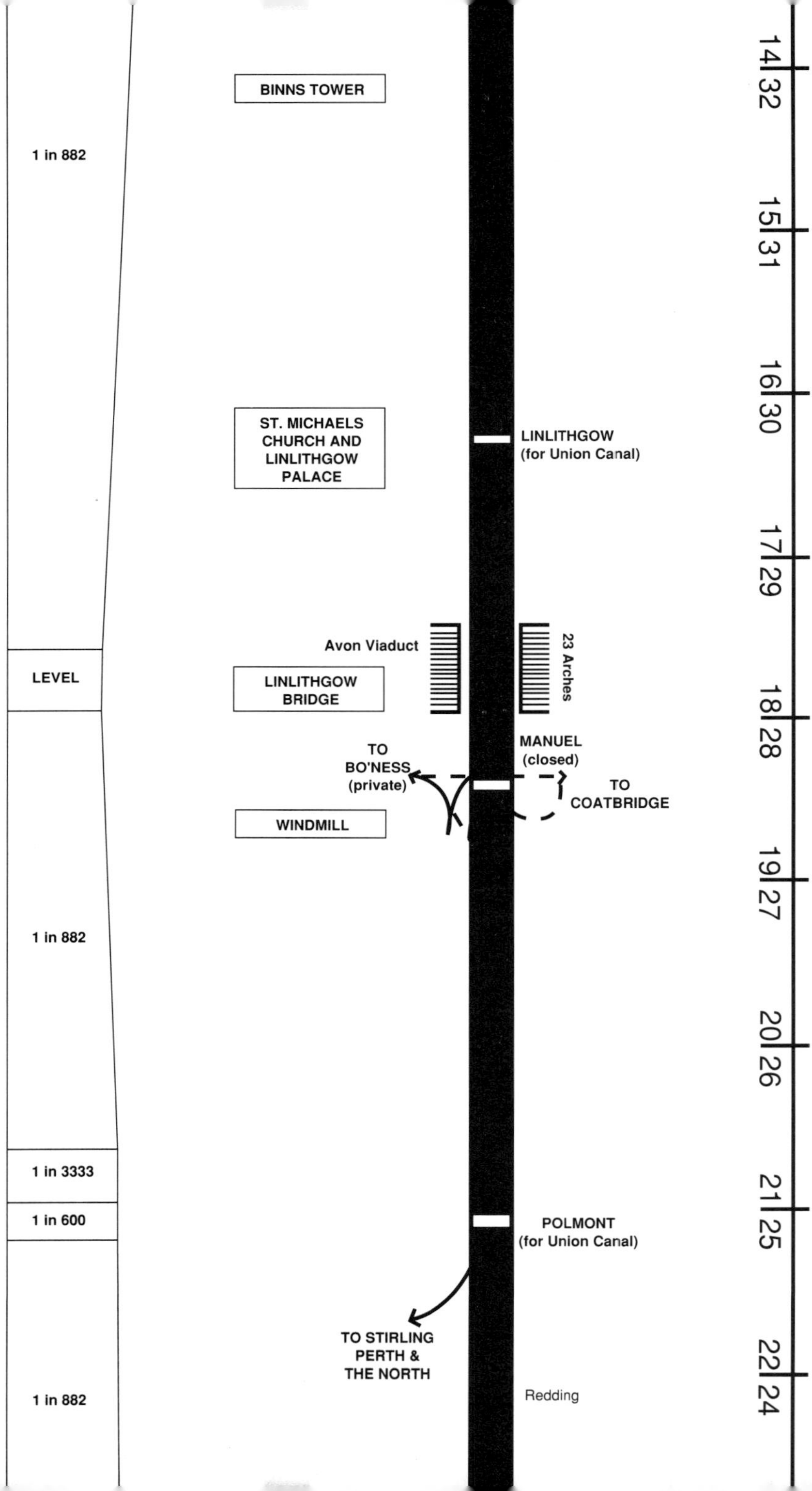

6. Linlithgow – Winchburgh

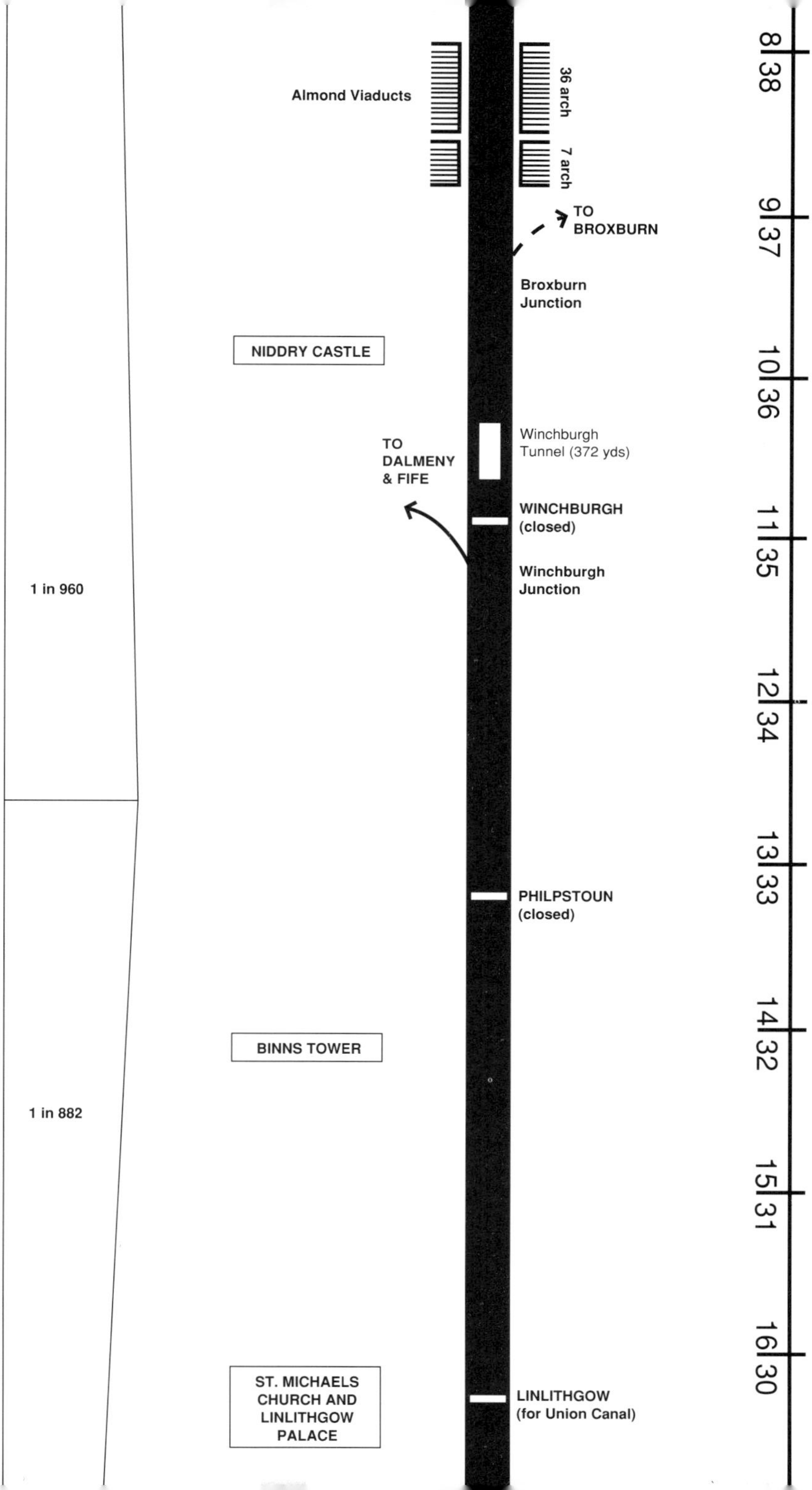

7. Winchburgh – Gogar

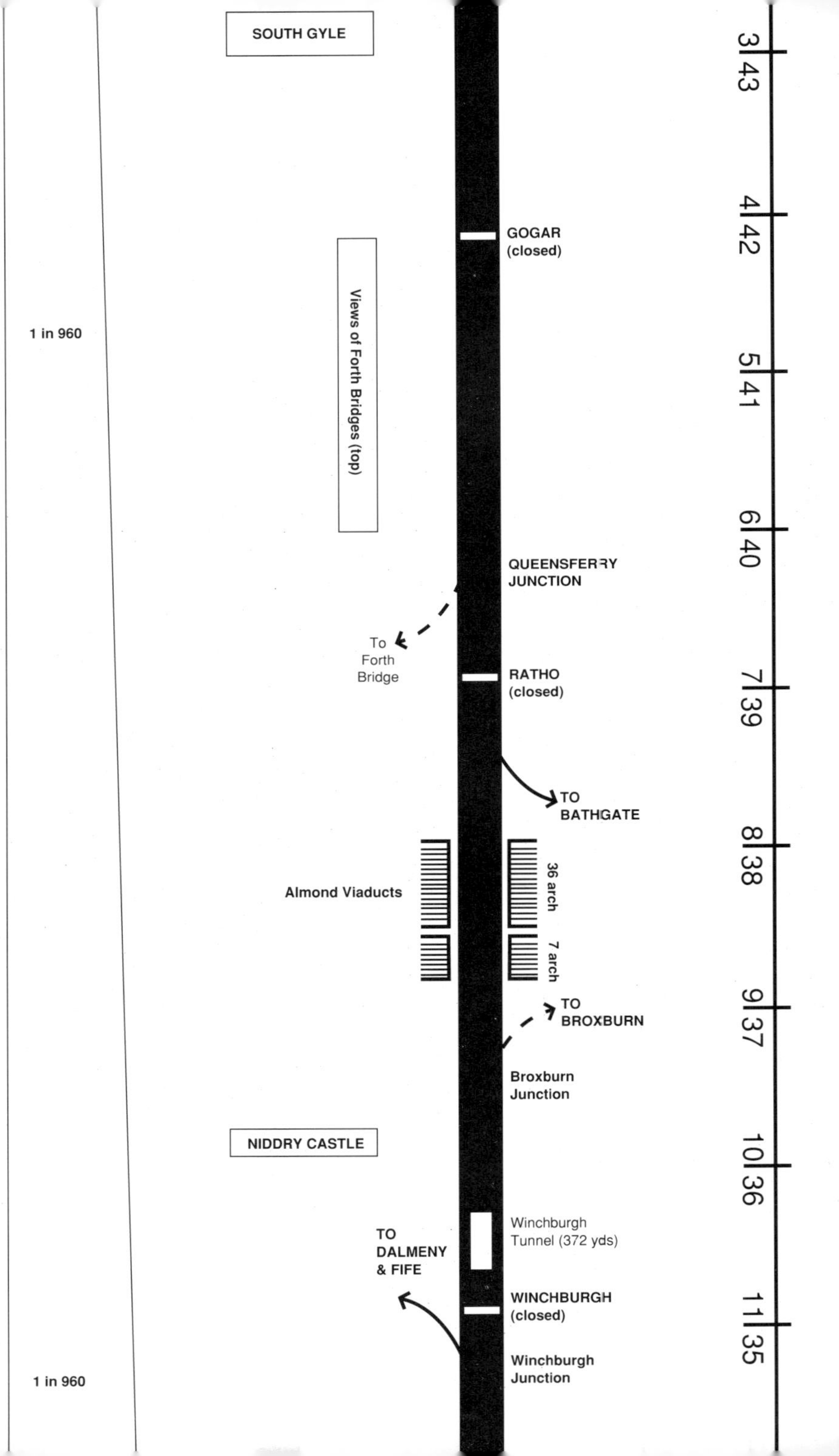

8. Gogar – Edinburgh

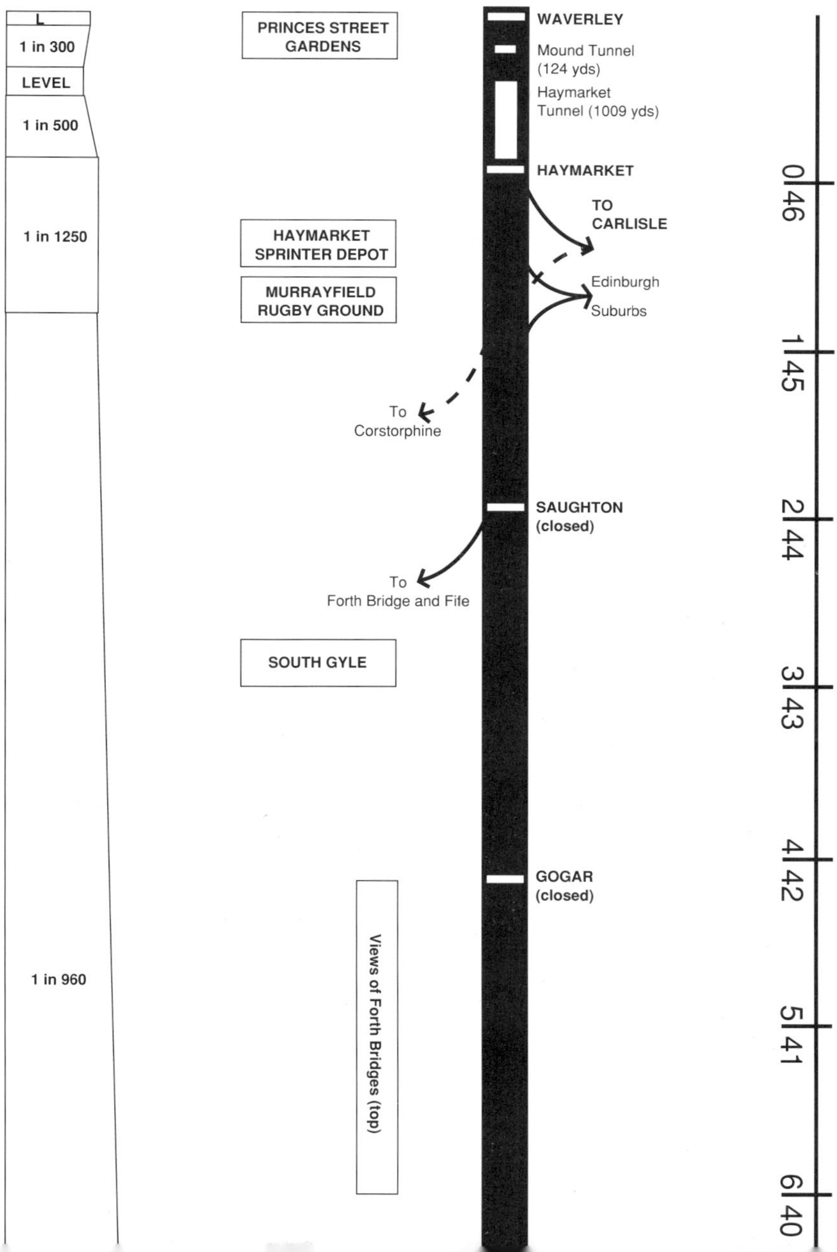

Bibliography

Railway History

Calder, J. — 'Notable railway stations, XII: Queen Street (High and Low Level), Glasgow', *Railway Magazine,* Vol. 8 (1901), pp. 145-154.

Collins, Michael J. — '20 years of Glasgow-Edinburgh push pulls', *Rail,* No.143 (March 1991), p.19.

Ellis, C. Hamilton — *The North British Railway* (1955).

Gairns, J.F. — 'Notable railway stations and their traffic: Queen Street, Glasgow, London and North Eastern Railway', *Railway Magazine*, Vol. 66 (1930), pp. 1-13.

Greville, Maurice D. — 'From Glasgow to Aberfoyle', *Railway Magazine,* Vol. 97 (1951), pp. 577-580.

Handley, James E. — *The Navvy in Scotland* (1970).

Johnston, Colin & Hume, John R. — *Glasgow stations* (1979).

Lockyer, A.E. — 'The Cowlairs incline on the North British Railway', *Railway World,* Vol. 3 (1894), pp. 343-348.

Maclean, A.A. — *North British album* (1975).

McLean, Allan P. — *This magnificent line: the story of the Edinburgh & Glasgow Railway* (1986).

Maley, Willy — 'The trial and execution of Dennis Doolan and Patrick Redding, Irish labourers on the Glasgow-Edinburgh railway line, 1841', *Scottish Labour History Review*, No.5 (1991-1992), pp. 7-9.

Moffat, Tom — 'Swing and other opening bridges on Scottish railways', *Railway Magazine*, Vol.28 (1911), pp. 389-396.

'Nimrod' — 'The Edinburgh and Glasgow Railway', *Railway Magazine*, Vol.6 (1900), pp. 37-45.

Robertson, C.J.A. — 'Early Scottish railways and the observance of the Sabbath', *Scottish Historical Review,* Vol.57 (1978), pp.143-167.

Robertson, C.J.A. — *The origins of the Scottish railway system, 1722-1844* (1983).

Robin, George H. — 'The Larbert-Polmont line', *Trains Illustrated,* Vol.6 (1953), pp. 56-59.

Simmons, Jack — *The railways of Britain* . 3rd ed. (1986).

Stoton, Frederick — 'North British Railway's Edinburgh and Glasgow main line', *Railway Magazine*, Vol.44 (1919), pp. 1-10.

Thomas, John — *Scotland, the Lowlands and the Borders, a regional history of the railways of Great Britain, VI.* 2nd ed. revised and enlarged by A.J.S. Paterson (1984).

Thomas, John	*The Springburn story* (1964).
Whishaw, Francis	*The railways of Great Britain and Ireland.* 2nd ed. (1842).
Willox, John	*Guide to the Edinburgh and Glasgow Railway* (1842).

Local History

Anderson, Rev.Robert	*A history of Kilsyth* (1901).
Bowman, A.I.	*Swifts & Queens: passenger transport on the Forth & Clyde Canal* (1984).
Butt, John	*The industrial archaeology of Scotland* (1967).
Campbell, R.H.	*Carron Company* (1961).
Carter, Paul, ed.	*Forth & Clyde Canal Guidebook.* 2nd ed. (1991).
Hendrie, William F.	*Linlithgow: six hundred years a royal burgh* (1989).
Hume, John R.	*The industrial archaeology of Scotland.* Vol.1 'The Lowlands and the Borders' (1976).
Hutton, Guthrie	*A Forth and Clyde canalbum* (1991).
Johnstone, Anne	*The wild frontier: exploring the Antonine Wall* (1986).
Lawson, Lewis	*A history of Falkirk* (1975).
Lindsay, Jean	*The canals of Scotland* (1968).
MacGibbon, David & Ross, Thomas	*The castellated and domestic architecture of Scotland.* 5 vols. (1887-92).
McGill, A.W & Preston W.	*Falkirk: short illustrated historical trails of Falkirk and surrounding area* (1984).
Martin, Don	*The story of Lenzie* (1989).
Massey, Alison	*The Edinburgh and Glasgow Union Canal* (1983).
Mulholland, Loretta	*Campsie Glen picture album* (1988).
Powell, Mark N.	*Linlithgow: a brief architectural and historical guide.* 2nd ed. (1990).
Pratt, Edwin A.	*Scottish canals and waterways* (1922).
Pringle, Denys	*Linlithgow Palace: a historical guide to the royal palace and peel* (1989). HMSO.
Robertson, Anne S.	*The Antonine Wall: a handbook to the surviving remains.* 4th ed. revised and edited by Lawrence Keppie (1990).
Robertson, David	*The Princes Street proprietors* (1935).
Sanderson, Kenneth W.	*The Scottish refractory industry, 1830-1980* (1990)
Simpson, Anne T. & Stevenson, Sylvia	*Historic Linlithgow: the archaeological implications of development* (1981).
Skinner, Basil C.	*The Union Canal: a short story of the Edinburgh & Glasgow Union Canal* (1977).

Royal Commission on Ancient and Historical Monuments & Constructions of Scotland. *Tenth report, with inventory of monuments and constructions in the counties of Midlothian and West Lothian* (1929).

Royal Commission on the Ancient and Historical Monuments of Scotland. *Stirlingshire.* 2 vols. (1963).

Acknowledgements

Strathkelvin District Libraries wish to acknowledge the extensive assistance they have received from ScotRail during preparation of this volume. In particular, Willie Caldwell of the Public Affairs Department has been most helpful, while the ScotRail official photographer, Gordon Smith, has been responsible for many of the photographs. Other photographs have been prepared by Eugene Duffy of Strathkelvin District Council Environmental Services Department. Rona Campbell of Strathkelvin District Libraries has word processed the manuscript, while the frontispiece map is the work of Stephen Moles of Strathkelvin District Press & Publicity Dept.

The authors wish to thank the various individuals who have been kind enough to submit photographs for publication, especially Douglas Hume, Tom Lindsay, Robin Nelson, Cecil Sanderson and Stuart Sellar. Suitable photographs from the collection of the late C. Lawson Kerr were identified by staff of the Glasgow Room at the Mitchell Library, where Ian Gordon was especially helpful. Others who have given assistance, in one way or another, include George Barbour of the Scottish Record Office, Edinburgh (where the official Edinburgh & Glasgow Railway records are located), Sheena McDougall of the Edinburgh Room at Edinburgh Central Library, Dr. Duncan Thomson of the Scottish National Portrait Gallery, Susan Dean and Helen Rowbotham of British Waterways, G. McVinnie of John Menzies UK Ltd and Tom Noble of Bishopbriggs. Special tribute is paid to the memory of the late James F. McEwan, one of the great fountainheads of Scottish railway knowledge, who has helped the authors with many projects over the years, including the present one up until his recent death. He will be sadly missed.

The book has been prepared for the 150th Anniversary of the opening of the Edinburgh & Glasgow Railway, 18th February 1992.

Don Martin
A.A. Maclean